OFFICIAL SQA PAST PAPERS

WITH ANSWERS

ADVANCED HIGHER

BIOLOGY
2008-2012

First exam published in 2008.
Published by Bright Red Publishing Ltd, 6 Stafford Street, Edinburgh EH3 7AU
tel: 0131 220 5804 fax: 0131 220 6710 info@brightredpublishing.co.uk www.brightredpublishing.co.uk

ISBN 978-1-84948-300-1

A CIP Catalogue record for this book is available from the British Library.

Bright Red Publishing is grateful to the copyright holders, as credited on the final page of the Question Section, for permission to use their material. Every effort has been made to trace the copyright holders and to obtain their permission for the use of copyright material. Bright Red Publishing will be happy to receive information allowing us to rectify any error or omission in future editions.

ADVANCED HIGHER

2008

[BLANK PAGE]

X007/701

NATIONAL
QUALIFICATIONS
2008

TUESDAY, 27 MAY
1.00 PM – 3.30 PM

BIOLOGY
ADVANCED HIGHER

SECTION A–Questions 1—25 (25 marks)

Instructions for completion of Section A are given on *Page two*.

SECTIONS B AND C

The answer to each question should be written in ink in the answer book provided. Any additional paper (if used) should be placed inside the front cover of the answer book.

Rough work should be scored through.

Section B (55 marks)

All questions should be attempted. Candidates should note that Question 8 contains a choice.

Question 1 is on Pages 10, 11 and 12. Questions 2, 3 and 4 are on Page 13. Pages 12 and 13 are fold-out pages.

Section C (20 marks)

Candidates should attempt the questions in **one** unit, **either** Biotechnology **or** Animal Behaviour **or** Physiology, Health and Exercise.

Read carefully

1   Check that the answer sheet provided is for **Biology Advanced Higher (Section A)**.

2   For this section of the examination you must use an **HB pencil** and, where necessary, an eraser.

3   Check that the answer sheet you have been given has **your name**, **date of birth**, **SCN** (Scottish Candidate Number) and **Centre Name** printed on it.

Do not change any of these details.

4   If any of this information is wrong, tell the Invigilator immediately.

5   If this information is correct, **print** your name and seat number in the boxes provided.

6   The answer to each question is **either** A, B, C or D. Decide what your answer is, then, using your pencil, put a horizontal line in the space provided (see sample question below).

7   There is **only one correct** answer to each question.

8   Any rough working should be done on the question paper or the rough working sheet, **not** on your answer sheet.

9   At the end of the exam, put the **answer sheet for Section A inside the front cover of the answer book**.

Sample Question

Which of the following molecules contains six carbon atoms?

A   Glucose

B   Pyruvic acid

C   Ribulose bisphosphate

D   Acetyl coenzyme A

The correct answer is **A**—Glucose. The answer **A** has been clearly marked in **pencil** with a horizontal line (see below).

Changing an answer

If you decide to change your answer, carefully erase your first answer and using your pencil, fill in the answer you want. The answer below has been changed to **D**.

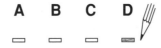

SECTION A

All questions in this section should be attempted.

Answers should be given on the separate answer sheet provided.

1. Which of the following genes encode proteins that promote normal cell division?

   A   Oncogenes

   B   Regulatory genes

   C   Proliferation genes

   D   Anti-proliferation genes

2. The cell cycle is believed to be monitored at checkpoints where specific conditions must be met for the cycle to continue.

   Condition 1:   chromosome alignment

   Condition 2:   successful DNA replication

   Condition 3:   cell size

   Which line in the table correctly shows the condition fulfilled at each checkpoint?

	G1	G2	M
A	2	3	1
B	3	1	2
C	2	1	3
D	3	2	1

3. Which line in the table correctly describes the chemical reaction that breaks down a disaccharide into its monomer subunits?

	Type of reaction	Type of bond broken
A	hydrolysis	peptide
B	condensation	glycosidic
C	hydrolysis	glycosidic
D	condensation	peptide

4. An unbranched polysaccharide is made up of glucose monomers joined together by β(1→4) linkages. The polysaccharide described could be

   A   amylose

   B   amylopectin

   C   glycogen

   D   cellulose.

5. Which of the following is a covalent bond that stabilises the tertiary structure of a protein?

   A   Hydrogen bond

   B   Disulphide bond

   C   Glycosidic bond

   D   Ester linkage

6. In the diagrams below, the sugar-phosphate backbone of a DNA strand is represented by a vertical line showing 5' to 3' polarity. The horizontal lines between bases represent hydrogen bonds.

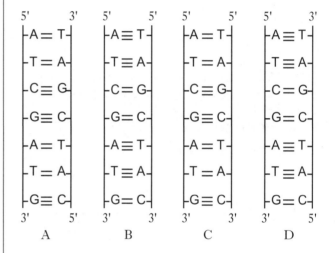

   Which diagram represents correctly a short stretch of a DNA molecule?

7. If ten percent of the bases in a molecule of DNA are adenine, what is the ratio of adenine to guanine in the same molecule?

   A   1:1

   B   1:2

   C   1:3

   D   1:4

[Turn over

8. Which line in the table below correctly summarises the movement of sodium and potassium ions into and out of a cell by a sodium-potassium pump?

	Potassium ions	*Sodium ions*
A	in	out
B	in	in
C	out	in
D	out	out

9. Which of the following is a component of the cytoskeleton?

A   Phospholipid

B   Tubulin

C   Peptidoglycan

D   Glycoprotein

10. A length of DNA is cut into fragments by the restriction enzymes BamHI and EcoRI

▼ BamHI cut site

△ EcoRI cut site

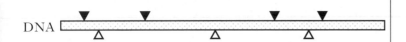

DNA

Which line in the table below correctly identifies the number of DNA fragments obtained?

	DNA cut by BamH1 only	*DNA cut by EcoR1 only*	*DNA cut by both BamH1 and EcoR1*
A	5	4	8
B	4	5	8
C	5	4	9
D	4	5	9

11. The diagram below outlines the stages involved in the polymerase chain reaction.

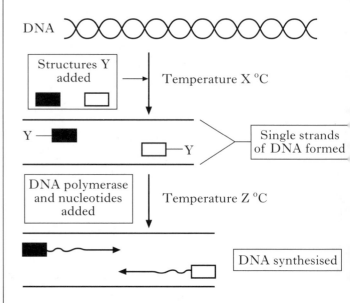

Which line in the table correctly identifies temperature X and the structures labelled Y?

	Temperature X (°C)	*Structure Y*
A	55	probe
B	95	primer
C	55	primer
D	95	probe

12. The result of profiling various DNA samples in a criminal investigation is shown below.

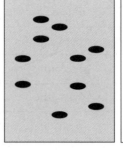

1 2 3 4 5

Key:
1   blood sample of victim
2   blood sample of suspect X
3   blood sample of suspect Y
4   first sample from forensic evidence
5   second sample from forensic evidence

Which of the following could the DNA analyst conclude about the crime?

A   Only suspect X was involved

B   Only suspect Y was involved

C   Suspects X and Y were both involved

D   Neither suspect X nor Y was involved

[Turn over for Section B on *Page ten*

SECTION B

All questions in this section should be attempted.
All answers must be written clearly and legibly in ink.

1. In 1992, a membrane protein called *aquaporin-1* (AQP1) was found to function exclusively as a channel for the passage of water molecules. The AQP1 molecule spans the phospholipid bilayer; each of its four linked subunits is a separate water channel (Figure 1).

Figure 1: Aquaporin-1 in a membrane

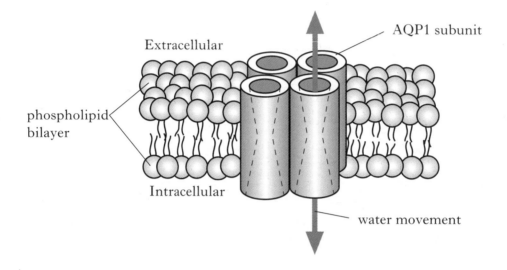

Early experiments were designed to determine the role of these protein channels in water movement. Researchers removed the contents of red blood cells to leave structures consisting of only the plasma membranes; these are known as "red cell ghosts". The ghosts were filled with solutions containing radioactive water and the concentration gradient across the membrane was varied. The rate at which water molecules moved out of the ghosts was measured in isotonic and hypertonic external solutions, before and after a treatment that inactivates AQP1.

From their results, summarised in the Table, the researchers concluded that the very rapid transfer of water across membranes during osmosis was through the AQP1 channels.

Table: Rate of water movement across ghost membranes

	Rate of water movement (units s^{-1})	
External solution	Untreated AQP1	Treated AQP1
Isotonic	2·5	1·0
Hypertonic	20	1·8

Recent studies have shown that several different aquaporins exist and they are present in a number of organs including eyes, brain, lungs and kidneys; all of them are important in water transport across membranes. In kidneys, they are found in some parts of the nephrons (Figure 2) where they have a role in water balance.

Question 1 (continued)

Figure 2: Nephron

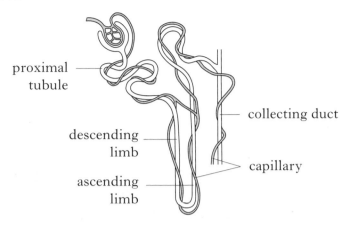

proximal tubule

collecting duct

descending limb

capillary

ascending limb

AQP1 is present only in the cells lining the proximal tubule, descending limb and in the capillaries associated with the nephron. About 70% of the water entering the nephron is reabsorbed here. A different aquaporin, AQP2, is present in cells of the collecting duct. The number of AQP2 molecules active at a cell surface increases with the concentration of antidiuretic hormone (ADH) in blood. ADH secretion is increased so that more water is reabsorbed into the capillaries from urine in the duct.

To study the importance of aquaporins in kidney function, three groups of mice with different genotypes were selected.

Group 1: genotype NN: homozygous for the presence of AQP1;

Group 2: genotype Nn: heterozygous;

Group 3: genotype nn: homozygous recessive; AQP1 absent.

Body mass and urine solute concentration were measured before and after a period without water. The results are shown in Figure 3 and Figure 4.

Figure 3: % change in body mass of mice after a period without water

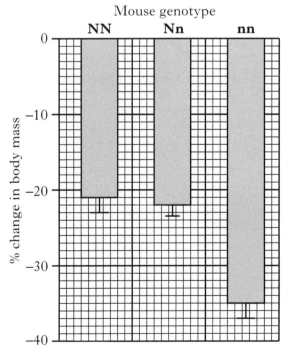

Figure 4: Urine solute concentration before (H) and after (D) a period without water

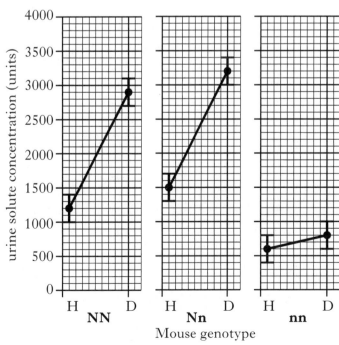

[Question 1 continues on *Page twelve*

Marks

Question 1 (continued)

(a)   (i)   What term describes a protein that is embedded in a membrane rather than attached to the surface?   1

(ii)   With reference to aquaporin, explain what is meant by the quaternary structure of a protein.   1

(b)   Refer to the experimental work using red cell ghosts.

(i)   Suggest an explanation for the use of radioactive water in the solutions placed in the ghosts.   1

(ii)   The Table shows that in isotonic conditions when AQP1 has been treated, water molecules flow out of the ghost cells at a rate of $1 \cdot 0$ units s^{-1}.

Explain why there would be no overall change in cell volume in these conditions.   1

(iii)   Aquaporins were inactivated by phosphorylation.

Which type of enzyme adds phosphate to a protein?   1

(iv)   Use data from the Table to show that functioning aquaporins can increase water flow across a membrane by over 1000%.   2

(c)   (i)   Figure 3 shows that homozygous recessive mice lost about 35% of their body mass during the period when they had no water supply.

Explain how these results may lead to the conclusion that the homozygous recessive mice lost abnormally high amounts of water in their urine.   2

(ii)   Refer to Figure 3 and Figure 4. Use the data to show that heterozygous mice are producing enough AQP1 molecules for normal osmoregulation.   3

(d)   Humans with a condition called *nephrogenic diabetes insipidus* (NDI) have normal AQP1 and ADH production but have non-functioning AQP2.

(i)   Predict the effect of a period of dehydration on urine production by individuals with NDI compared to individuals without NDI.   1

(ii)   Explain your prediction.   1

                                                                                                                                                (14)

[Questions 2, 3 and 4 are on fold-out *Page thirteen*

Marks

2. (*a*) (i) Identify structure X shown in the diagram below.   **1**

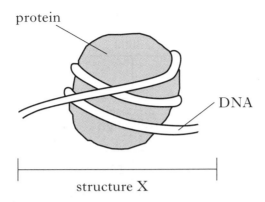

(ii) Why is the binding of DNA to this type of protein so important for eukaryotic cells?   **2**

(*b*) In the production of transgenic plants, the genome of a plant species can be modified by incubating protoplasts with engineered plasmids.

(i) Name the prokaryotic species used as the source of these plasmids.   **1**

(ii) Describe how protoplasts are produced from isolated plant cells.   **1**

     (5)

3. Describe the general structure of steroids and their function in cell signalling.   **(4)**

4. The diagram below represents an enzyme, PRPP synthetase, involved at the start of the biochemical pathway that produces nucleotides. In the active site there are two positions (S) where the substrate molecules, ribose and ATP, bind and react. Position I is an inhibitor binding site and position A is an activator binding site.

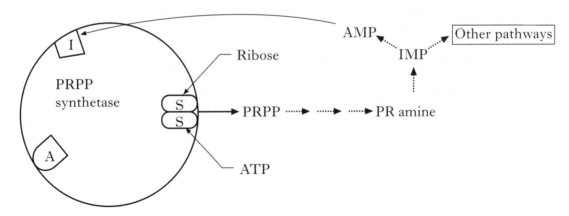

(*a*) What is meant by induced fit when referring to enzyme action?   **1**

(*b*) Explain why PRPP synthetase is described as an allosteric enzyme.   **1**

(*c*) Describe the effect of AMP formation on the metabolic pathway.   **2**

     (4)

Marks

5. The diagram below shows energy flow through a deciduous forest ecosystem. Units are $kJ\ m^{-2}\ day^{-1}$.

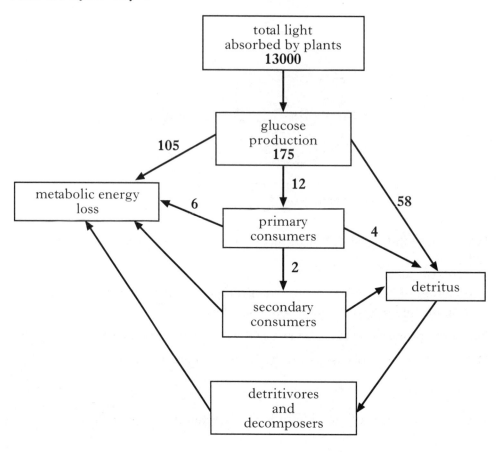

(a) Calculate the percentage of light energy that is captured in photosynthesis by the forest plants.   1

(b) What is the value for net primary productivity (NPP) in this ecosystem?   1

(c) Ten percent is often quoted as a typical value for ecological efficiency. Use the data to show that this value does not always apply to energy transfer between trophic levels.   1

(d) In what form is the energy lost from metabolism?   1

(e) Compare the use of digestive enzymes by detritivores and decomposers.   1

(5)

Marks

6. The figure below shows processes in the nitrogen cycle that rely on the activities of micro-organisms.

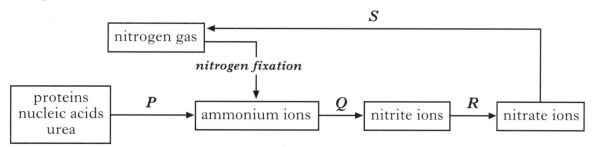

(a) The bacteria responsible for processes Q and R are *obligate aerobes*.

What does this term suggest about their metabolic requirements?   **1**

(b) Nitrogen fixation is often the product of a prokaryote–eukaryote symbiosis.
It depends on the action of an enzyme system that functions best in anaerobic conditions.

  (i)   Identify a prokaryote involved in such an interaction.   **1**

  (ii) Name the enzyme responsible for nitrogen fixation.   **1**

  (iii) What is the role of leghaemoglobin in this interaction?   **1**

(c) State the likely effect of process S on soil fertility in aerobic conditions.   **1**

(5)

7. Rust fungi are pathogens of many monocultures. In infected plants, *pustules* are formed that are responsible for the spread of infection.

The figure below shows the spread of a rust fungus from a single infected plant.

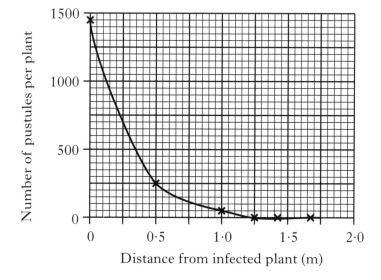

(a) Define the term monoculture.   **1**

(b) What do the data suggest about the spacing needed to prevent the fungus infection from spreading?   **1**

(c) Apart from altering spacing, suggest another way of growing rust-free crops.   **1**

(3)

Marks

8. Answer **either** A **or** B.

    A. Discuss how ecosystems may be affected by the following:

        (i) phosphate enrichment; **5**

        (ii) exotic species; **4**

        (iii) persistent toxic pollutants. **6**

  OR **(15)**

    B. Discuss the roles of the following in the survival of organisms:

        (i) dormancy; **7**

        (ii) mimicry; **4**

        (iii) mutualism. **4**

                                                (15)

[END OF SECTION B]

SECTION C

Candidates should attempt questions on <u>one</u> unit, <u>either</u> Biotechnology <u>or</u> Animal Behaviour <u>or</u> Physiology, Health and Exercise.

The questions on Animal Behaviour can be found on pages 20–22.

The questions on Physiology, Health and Exercise can be found on pages 23–25.

All answers must be written clearly and legibly in ink.

Labelled diagrams may be used where appropriate.

Marks

Biotechnology

1. (a) The Figure below shows the growth curve of a bacterium.

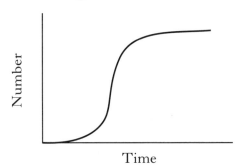

Give **two** reasons why there is only a small change in the number of bacteria during the lag phase.

2

   (b) A haemocytometer is used to estimate cell numbers. The diagram below shows part of a haemocytometer grid. The depth of the chamber is 0·1 mm.

North

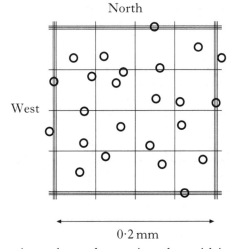

0·2 mm

   (i) One precaution taken when using the grid is to exclude cells overlapping the north and west sides.

What is the purpose of this precaution?

1

   (ii) Calculate the number of cells in 1 mm³ of the sample shown on the haemocytometer grid.

1

   (c) (i) Penicillin is an antibiotic that is described as *bacteriostatic*.

Distinguish between *bacteriostatic* activity and *bactericidal* activity.

1

   (ii) Name an antibiotic other than penicillin.

1

   (d) Antibodies are produced in response to the presence of foreign antigens.

   (i) Name the cells that secrete antibodies.

1

   (ii) Give **one** medical use of monoclonal antibodies.

1

(8)

Biotechnology (**continued**)   *Marks*

2. Explain how genetic modification of the "*flavrsavr*" tomato plant has resulted in fruit with longer shelf life.   **(4)**

3. (*a*) The nitrogen content of two yeast extracts, BW6 and BK1, is shown in the Table below.

	Yeast extract	
	BW6	*BK1*
Total nitrogen (g/100 g)	7·1	11·9
% of nitrogen as amino acids	34	56

The figure below shows the effect on the growth of the bacterium *Lactobacillus casei* of adding different proportions of BW6 and BK1.

Figure: Effect of varying proportions of BW6 and BK1 on the growth of *L. casei*

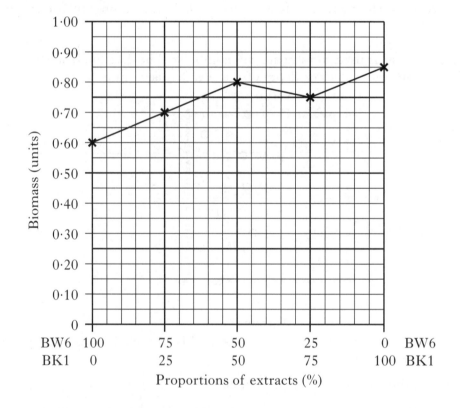

   (i) Use the data to describe the effect of varying the proportions of the extracts on the growth of the bacterium.   2

   (ii) Suggest a reason for this effect on *L. casei*.   1

(*b*) Give one agricultural use of *Lactobacillus* species.   1

   (4)

Biotechnology **(continued)**   *Marks*

4. Different yeast species are cultured to create large quantities of yeast cell biomass. Some of the biomass is destined for activities such as brewing and baking while much of the biomass is treated to bring about autolysis. The *autolysate* produced has a variety of uses, for example as a nutrient source in fermentation media.

    (*a*) What is meant by the term *autolysis*?   1

    (*b*)   State **two** factors that could influence the characteristics of yeast autolysate.   2

    (*c*) Yeast extract is prepared from the autolysate.

        State one use of yeast extract in the food industry.   1

                                                                             (4)

                                                                             (20)

[End of *Biotechnology* questions. *Animal Behaviour* questions start on Page 20]

[Turn over

SECTION C (continued)   *Marks*

Animal Behaviour

1. The woodlouse *Hemilepistus reaumuri* lives in deserts. To avoid drying out, it digs a burrow from which it ventures each day to forage.

    Figure: Life cycle of *H. reaumuri*

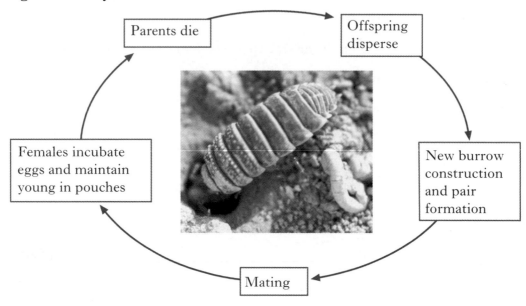

    (*a*) After dispersal, both males and females can begin new burrows, or they can attempt to pair with an existing burrower.

    Why are larger males more successful than smaller ones in pair formation with a female that has already started a burrow?   1

    (*b*) Suggest proximate and ultimate causes for burrow digging in this species.   2

    (*c*) For *H. reaumuri*, state one reason why the parental investment of females is higher than that of males.   1

    (*d*) After foraging, these woodlice show an *innate* ability to navigate straight back to their burrows.

    What is meant by the term innate?   1

    (*e*) State the effect of an increased encounter rate and a reduced handling time on the duration of foraging.   1

                                                                                       (6)

Animal Behaviour **(continued)** *Marks*

2. The fire ant *Solenopsis invicta* is a social insect found in South America. A study of kin selection showed that colonies of *S. invicta* can have either of two distinct social structures, *compact* colonies or *sprawling* colonies.

   Compact colonies contain ants that are all produced by a single female, the queen. The workers in compact colonies are loyal to the queen and are aggressive towards intruders. In contrast, large sprawling colonies, which are becoming more common, are formed from many interconnected nests inhabited by many queens.

   The difference in social structure depends on the presence or absence of allele B of the gene *GP-9*. Ants with the B allele produce a receptor protein that enables them to distinguish the odours of ants from different genetic backgrounds.

   (a) What is meant by the term *kin selection*? 1

   (b) Allele B is only found in compact colonies.

   Explain how the absence of allele B has led to the formation of the sprawling colony. 2

   (c) Give another example of a single gene effect. 1

   (4)

3. Discuss the role of dispersal in the avoidance of inbreeding in birds and mammals.

   Why is it important for animals to avoid inbreeding and how is this achieved? **(4)**

[Turn over

Animal Behaviour **(continued)**   *Marks*

4. The success of herring gull (*Larus argentatus*) foraging behaviour was studied using video recordings. Herring gulls have adult plumage after 4 years; the age of younger birds can be determined by other visible characteristics.

    The Figure shows the mean feeding rates for herring gulls in different age classes.

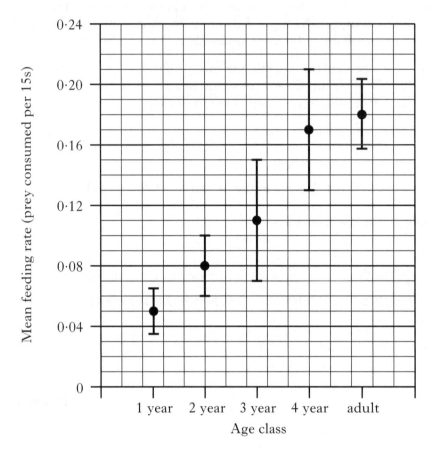

    (*a*)   (i)   Identify **two** age classes between which there is a significant difference in foraging success.   **1**

         (ii)   Suggest an explanation for the trend seen in the feeding rate.   **1**

         (iii)   An adult feeding at 0·18 prey per 15 seconds takes 83 seconds to capture one prey item.

              How long would it take a one-year-old to capture one prey item?   **1**

    (*b*)   Suggest **one** advantage of using video recording for this analysis.   **1**

    (*c*)   Describe **two** ways in which herring gulls have successfully adapted to human influence.   **2**

                                                            (6)

                                                            (20)

[End of *Animal Behaviour* questions. *Physiology, Health and Exercise* questions start on Page 23]

SECTION C (continued)   *Marks*

Physiology, Health and Exercise

1.   A stent is a narrow, wire mesh tube that can be inserted into a blood vessel. It may be used to treat atherosclerosis in the blood vessels of the heart. Figure 1 shows a blood vessel before and after the procedure to insert a stent.

Figure 1

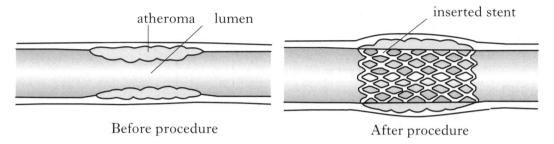

atheroma   lumen   inserted stent

Before procedure   After procedure

(*a*) Name the blood vessels that deliver blood to the myocardium.   **1**

(*b*) Describe how *atherosclerosis* develops.   **2**

(*c*) Figure 2 shows the volume changes in a blood vessel in the heart as a result of a stent being fitted.

Figure 2

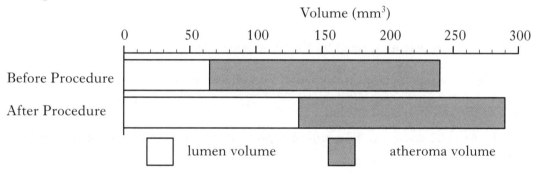

Volume (mm^3)

lumen volume   atheroma volume

(i)   Use the data to explain how the procedure achieves increased blood flow.   **2**

(ii)   What term is used for the chest pain relieved by this procedure?   **1**

                                              (6)

[Turn over

Physiology, Health and Exercise **(continued)**   *Marks*

2. The data show trends expected in the numbers of American women who have low bone mass and who will go on to develop osteoporosis.

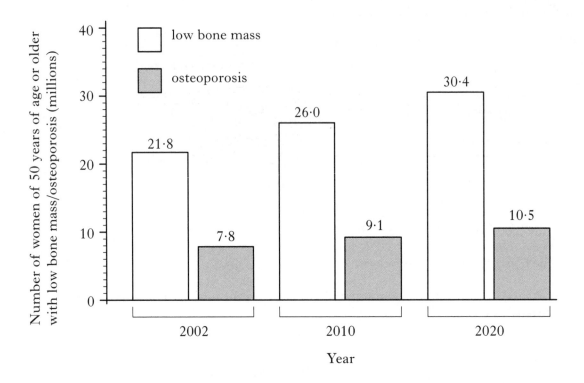

(a) Give one feature of osteoporosis other than low bone mass.   **1**

(b) Explain why the study focuses on women of 50 years of age or older.   **1**

(c) Authors of the research claimed that:

"*In future, fewer American women with hip problems will eventually go on to develop osteoporosis.*"

(i) Suggest how critics could use the data provided to contradict this claim.   **1**

(ii) How did the authors use the data to arrive at their conclusion?   **1**

(d) Explain why American teenagers would be advised to take up jogging rather than swimming to reduce the risk of osteoporosis.   **2**

                                                                                                (6)

Physiology, Health and Exercise **(continued)**   *Marks*

3. (*a*) The table shows data relating to four members of a group trying to achieve different weight-loss targets. The energy deficit value indicates the severity of their intended diet. Dietary weight loss is assumed to be from fat loss.

A negative energy balance of 29·4 MJ is required to lose 1 kg.

Group member	Present weight (kg)	Target weight (kg)	Energy deficit (MJ/day)
A	96	91	2
B	121	113	3
C	104	98	2
D	92	82	3

   (i)   How many days will individual **D** take to reach the target weight?   1

   (ii) Individual **A** is 1·74m in height. Calculate this individual's body mass index (BMI).   1

   (*b*) Explain why being "overweight" does not always mean that someone is unhealthy.   1

   (*c*) Name a method of measuring body composition.   1

                                                                                     (4)

4. Discuss the use of exercise testing in the assessment of aerobic fitness.   **(4)**

                                                                                     (20)

[END OF QUESTION PAPER]

[BLANK PAGE]

[BLANK PAGE]

X007/701

NATIONAL
QUALIFICATIONS
2009

THURSDAY, 28 MAY
1.00 PM – 3.30 PM

BIOLOGY
ADVANCED HIGHER

SECTION A–Questions 1—25 (25 marks)

Instructions for completion of Section A are given on *Page two*.

SECTIONS B AND C

The answer to each question should be written in ink in the answer book provided. Any additional paper (if used) should be placed inside the front cover of the answer book.

Rough work should be scored through.

Section B (55 marks)

All questions should be attempted. Candidates should note that Question 8 contains a choice.

Question 1 is on Pages 10, 11 and 12. Questions 2, 3 and 4 are on Page 13. Pages 12 and 13 are fold-out pages.

Section C (20 marks)

Candidates should attempt the questions in **one** unit, **either** Biotechnology **or** Animal Behaviour **or** Physiology, Health and Exercise.

Read carefully

1   Check that the answer sheet provided is for **Biology Advanced Higher (Section A)**.

2   For this section of the examination you must use an **HB pencil** and, where necessary, an eraser.

3   Check that the answer sheet you have been given has **your name**, **date of birth**, **SCN** (Scottish Candidate Number) and **Centre Name** printed on it.

    Do not change any of these details.

4   If any of this information is wrong, tell the Invigilator immediately.

5   If this information is correct, **print** your name and seat number in the boxes provided.

6   The answer to each question is **either** A, B, C or D. Decide what your answer is, then, using your pencil, put a horizontal line in the space provided (see sample question below).

7   There is **only one correct** answer to each question.

8   Any rough working should be done on the question paper or the rough working sheet, **not** on your answer sheet.

9   At the end of the exam, put the **answer sheet for Section A inside the front cover of the answer book**.

Sample Question

Which of the following molecules contains six carbon atoms?

A   Glucose

B   Pyruvic acid

C   Ribulose bisphosphate

D   Acetyl coenzyme A

The correct answer is **A**—Glucose. The answer **A** has been clearly marked in **pencil** with a horizontal line (see below).

Changing an answer

If you decide to change your answer, carefully erase your first answer and using your pencil, fill in the answer you want. The answer below has been changed to **D**.

SECTION A

All questions in this section should be attempted.

Answers should be given on the separate answer sheet provided.

1. The diagram shows a bacterial cell.

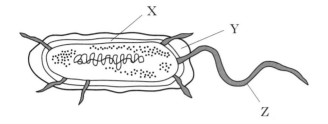

Which line in the table below correctly identifies the labelled structures?

	X	Y	Z
A	cell wall	capsule	flagellum
B	capsule	cell wall	flagellum
C	cell wall	capsule	pilus
D	capsule	cell wall	pilus

2. Which of the following diagrams best represents the sequence of phases involved in the cell cycle?

A

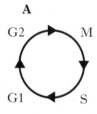

B

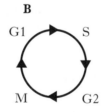

C

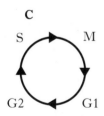

D

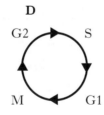

3. The covalent chemical bonds between nucleotides in DNA are

A peptide

B phosphodiester

C glycosidic

D hydrogen.

4. Which line in the table below classifies correctly the four bases in DNA as either purines or pyrimidines?

	Purines	Pyrimidines
A	adenine and thymine	cytosine and guanine
B	cytosine and guanine	adenine and thymine
C	cytosine and thymine	adenine and guanine
D	adenine and guanine	cytosine and thymine

5. The table below shows the number of cells from a cell culture at different points in the cell cycle.

Stage	Number of cells
Interphase	462
Prophase	23
Metaphase	24
Anaphase	4
Telophase	16

The mitotic index of the sample is

A 14·5%

B 12·7%

C 85·5%

D 74·7%.

[Turn over

6. The percentage of adenine bases in a double stranded DNA molecule is 30% and in a single stranded RNA molecule it is 25%.

Which line in the table below shows the number of other bases in each molecule for which the percentage could be calculated?

	RNA	DNA
A	none	three
B	none	none
C	one	two
D	one	three

7. During a biochemical reaction the transfer of a phosphate group from one molecule to another is catalysed by

A   ligase

B   ATPase

C   kinase

D   nuclease.

8. The diagram below shows an enzyme-catalysed reaction.

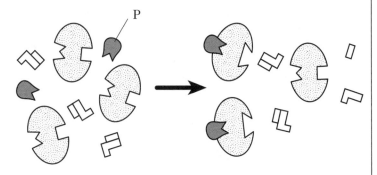

Which of the following correctly identifies molecule P?

A   Substrate

B   Activator

C   Competitive inhibitor

D   Non-competitive inhibitor

9. The diagram below shows the changes in the activity of enzymes that synthesise tryptophan and utilise lactose in a cell after the addition of tryptophan and lactose.

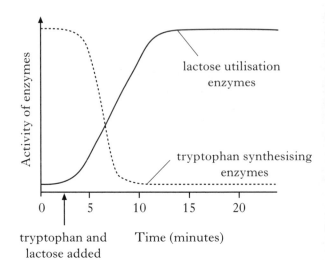

What valid conclusion may be made from the graph?

A   Addition of lactose acts as a negative enzyme modulator.

B   Addition of tryptophan acts as a positive enzyme modulator.

C   Enzyme induction is occurring in lactose utilisation enzymes.

D   Enzyme induction is occurring in tryptophan synthesising enzymes.

10. During the production of transgenic plants, which of the following bacteria would be used to transfer recombinant plasmids into plant protoplasts?

A   *Agrobacterium*

B   *E. coli*

C   *Pseudomonas*

D   *Rhizobium*

11. The diagram below shows the restriction enzyme sites in a plasmid that carries the genes for resistance to the antibiotics ampicillin and tetracycline.

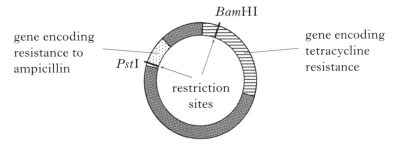

Which line in the table below identifies correctly the antibiotic resistance that would remain when a gene is inserted at these restriction enzyme sites?

	Gene inserted into restriction enzyme site	*Antibiotic resistance remaining*
A	*Bam*HI	tetracycline and ampicillin
B	*Pst*I	ampicillin
C	*Pst*I	tetracycline and ampicillin
D	*Bam*HI	ampicillin

12. A piece of DNA was digested using the restriction enzymes *Bam*HI and *Eco*RI. The results are shown below.

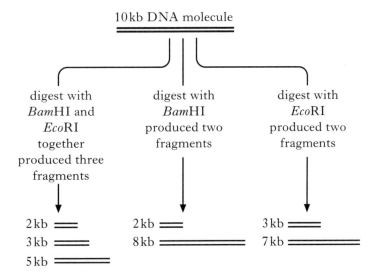

Which of the following restriction maps can be drawn from these results?

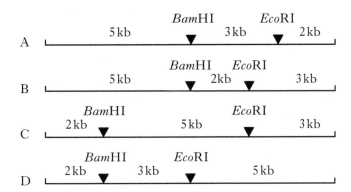

13. The graph below shows variation in biomass throughout one year in an aquatic ecosystem.

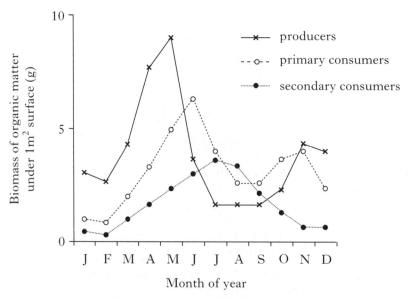

During which month of the year would the following pyramid of biomass be applicable?

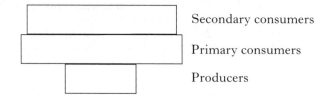

Secondary consumers

Primary consumers

Producers

A   June

B   July

C   August

D   September

14. The table below shows measurements of energy in a grassland ecosystem.

	Units of energy m^{-2} $year^{-1}$
Solar energy entering ecosystem	471·00
Fixed in photosynthesis	5·83
Released in respiration by autotrophs	0·88

What is the net productivity (*units of energy* m^{-2} $year^{-1}$) for this ecosystem?

A   4·95

B   6·71

C   465·17

D   470·12

15. Which of the following statements best describes a detritivore?

A   Micro-organism with external enzymatic digestion

B   Micro-organism with internal enzymatic digestion

C   Invertebrate with external enzymatic digestion

D   Invertebrate with internal enzymatic digestion

16. The release of nutrients from the remains of dead organisms in the soil is called

A   assimilation

B   humus formation

C   mineralisation

D   nitrification.

17. Which of the following promotes the loss of nitrogen from soil due to the activity of denitrifying bacteria?

   A   Leaching of nitrate from soil in drainage water

   B   Anaerobic conditions caused by water saturation

   C   High levels of phosphate from addition of fertilisers

   D   The presence of a leguminous crop such as clover

18. A flask containing a solution of ammonium salts was set up to demonstrate the activity of some of the micro-organisms involved in the nitrogen cycle. A sample of fresh soil was added to the solution and the concentration of nitrite measured over several weeks. The results are shown in the graph below.

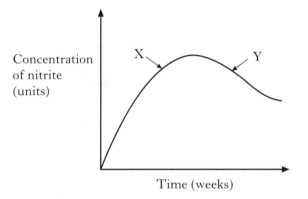

Concentration of nitrite (units)

Time (weeks)

Which line in the table below correctly represents bacterial activity which can account for the changes shown at X and Y?

	Bacteria active at X	*Bacteria active at Y*
A	*Nitrosomonas*	*Rhizobium*
B	*Nitrobacter*	*Nitrosomonas*
C	*Nitrobacter*	*Rhizobium*
D	*Nitrosomonas*	*Nitrobacter*

19. Coral snakes are highly venomous and have a pattern of dark red, yellow and black bands.

   This is an example of

   A   aposematic colouration

   B   Batesian mimicry

   C   camouflage

   D   Mullerian mimicry.

20. A species of Latin American ant inhabits the thorns of a species of *Acacia*. The ant receives nectar and shelter from the plant. The plant receives protection from the ants.

   This is an example of

   A   parasitism

   B   commensalism

   C   mutualism

   D   predation.

21. *Hydra* is a small freshwater animal that uses its tentacles to catch food. One variety (green hydra) has photosynthetic algae living in its tissues. Another variety (colourless hydra) has no algae.

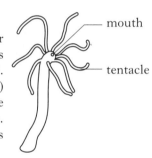

mouth

tentacle

   The relationship between *Hydra* and the algae is believed to be an example of mutualism.

   Under what conditions would a comparison of the growth rates of green and colourless *Hydra* test this hypothesis?

   A   Light; food supplied

   B   Light; no food supplied

   C   Dark; food supplied

   D   Dark; no food supplied.

[Turn over

22. Animals may interact with their environment by conformation or regulation. Each statement below applies to one of these interactions.

1 A wide range of habitats can be occupied.
2 A restricted range of habitats can be occupied.
3 There is a high energy cost.
4 There is no energy cost.

Which statements apply to regulation?

A   1 and 3 only

B   1 and 4 only

C   2 and 3 only

D   2 and 4 only

23. The graph below shows primary productivity in a loch at different depths. Data were collected before and after an experiment in which phosphate was added to the loch.

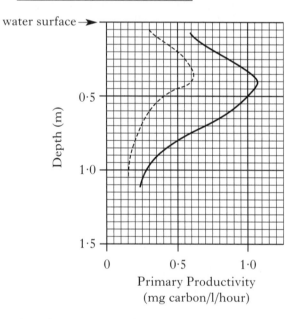

Calculate the percentage increase in productivity at a depth of 0·5 m that results from the addition of phosphate.

A   60%

B   75%

C   150%

D   300%

24. Which line in the table correctly identifies the effect of each pollutant?

	Biomagnification	Eutrophication	Global warming
A	phosphate	DDT	CFCs
B	DDT	phosphate	CFCs
C	CFCs	phosphate	DDT
D	DDT	CFCs	phosphate

25. The concentrations of some toxic organic chemicals in sea water were compared to concentrations known to produce lethal effects in laboratory experiments.

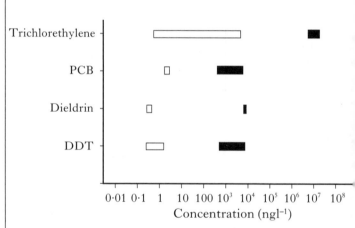

Which of the following is a valid conclusion from the data shown?

A   All the toxic organic chemicals are found at lethal concentrations in sea water.

B   Trichlorethylene is the only chemical found at lethal concentrations in sea water.

C   DDT produces toxic effects in sea water due to biomagnification through the ecosystem.

D   There is no evidence that the concentrations of toxic organic chemicals in sea water could produce lethal effects.

[END OF SECTION A]

Candidates are reminded that the answer sheet MUST be returned INSIDE the front cover of the answer book.

[Turn over for Section B on *Page ten*]

SECTION B

All questions in this section should be attempted.
All answers must be written clearly and legibly in ink.

1. The effects of large carnivores on ecosystems are not well understood. Predators can have a "top-down" effect that ripples down through the trophic levels below them. The effect is called a *trophic cascade*. In a trophic cascade a predatory species significantly affects consumer populations, which in turn results in significant changes at the producer level. Two recent studies in the USA have investigated trophic cascades.

    One study investigated the effects of reintroducing wolves (*Canis lupus*) to Yellowstone National Park in 1994. The strength of the trophic cascade was assessed by measuring feeding damage caused by elk (*Cervus elaphus*) to saplings produced by regenerating aspen trees (*Populus tremuloides*) (Table 1).

    In the second study, a trophic cascade was quantified in a comparison of two neighbouring canyons in Zion National Park. These two canyons, Zion Canyon and North Creek, have similar geology, climate and plant species but are visited by substantially different numbers of tourists. The Figure shows the age structure of populations of cottonwood trees (*Populus fremontii*) in the two canyons. The age of cottonwood trees was estimated by measuring their diameter at chest height. Other than the trees, the most significant species within the community are the predatory cougar (*Puma concolor*) and the herbivorous mule deer (*Odocoileus hemionus*).

    To compare the abundance of these species in the two canyons, data were collected from two-metre-wide transects following the course of river and stream banks. Evidence of cougars, which are highly sensitive to human disturbance, was determined by searching for scats (droppings) along 4000m of walking trails in the two localities (Table 2).

Table 1 : Survey data for Yellowstone National Park

Year	Wolf population	Elk population	Feeding damage (%)	Average aspen sapling height (cm)
1993	0	17 500	No data	No data
1997	24	13 000	95	30
2001	74	12 000	80	50
2005	82	9000	20	170

Question 1 (continued)

Figure: Age structure of cottonwood trees in two canyons in Zion National Park

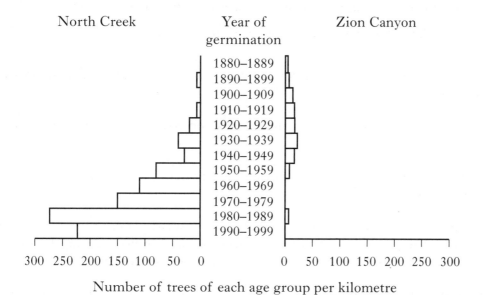

Number of trees of each age group per kilometre

Table 2: Comparative data for three species in Zion National Park

	Canyon	
Species	North Creek	Zion Canyon
Cougar (scats per km)	1·75	0
Deer (hoof prints per km)	3·3	700
Young cottonwood (saplings per km)	900	23

[Question 1 continues on *Page twelve*

Marks

Question 1 (continued)

(a) Top-down effects are caused by heterotrophs. What is meant by the term heterotroph?   1

(b) Using data from Table 1:

    (i)   describe the trophic cascade caused by the reintroduction of wolves;   2

    (ii) calculate the percentage increase in wolf population over the period 1997 to 2001;   1

    (iii) suggest **one** reason why the herbivore population declined by less than 8% over the same period.   1

(c) Compare the abundance of old and young cottonwood trees in North Creek and Zion Canyon.   2

(d) Suggest why the investigators used cougar scats rather than sightings to estimate cougar abundance.   1

(e) Zion Canyon has been accessible to a large number of tourists since the 1930s, whereas North Creek is rarely visited. Justify the conclusion that, by influencing the trophic cascade, tourism is responsible for the poor survival of young cottonwoods.   2

(f) (i)   What term describes biotic effects that increase in intensity as the population in an area increases?   1

    (ii) Explain how the intensity of grazing can influence the *diversity* of plant species.   2

(13)

[**Questions 2, 3 and 4 are on fold-out** *Page thirteen*

Marks

2. (a) Himalayan balsam (*Impatiens glandulifera*) is an exotic species that spread into Scotland after being introduced into the UK in 1839. Left unchecked it can form an ecologically harmful monoculture. Himalayan balsam is an annual plant (its whole life cycle takes place within one year). Seeds of this species can remain dormant in the soil for up to two years.

    (i)   What is the benefit of a period of dormancy in seeds?   **1**

    (ii) Describe a damaging effect arising from the spread of an exotic species.   **1**

    (iii) Suggest a method for controlling Himalayan balsam.   **1**

    (iv) How would the seed dormancy of the Himalayan balsam affect the design of an eradication programme?   **1**

(b) Give **one** effect on soil when monoculture is used in intensive food production.   **1**

    (5)

3. Explain how the use of fossil fuels disrupts the symbiotic relationship in coral.   **(4)**

4. (a) Why is competition regarded as a negative interaction?   **1**

(b) Explain what is meant by a fundamental niche.   **1**

(c) A survey of birds in the Bismarck Islands, Papua New Guinea, found that two similar species of cuckoo-doves, *Macropygia mackinlayi* and *M. nigrirostris*, are never found breeding on the same island.

     M. mackinlayi   *M. nigrirostris*

The two species are believed to have very similar fundamental niches. Suggest an explanation for the two species occupying different islands.   **2**

(d) Parasites may be transmitted between closely related species.

    (i)   Why is the transmission of parasites less common between **unrelated** species?   **1**

    (ii) State **one** way in which parasites can be transmitted.   **1**

    (6)

Marks

5. The diagram below shows a section of plasma membrane with proteins labelled A to E.

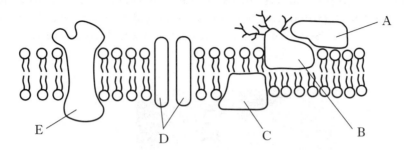

(a) (i)   Identify which of the proteins A to E are integral membrane proteins.   1

    (ii)   Which type of signalling molecule requires a receptor protein at the cell surface?   1

(b) The membranes of most eukaryotic cells contain a proportion of the steroid cholesterol.

    (i)   Describe the general structure of a steroid.   1

    (ii)   State **one** role of cholesterol in membranes.   1

   (iii)   The table below shows the proportion of cholesterol in membranes from different locations.

Membrane location	Proportion of cholesterol (g cholesterol per g membrane)
Liver plasma membrane	0·18
Mitochondrial membrane	0·03
Endoplasmic reticulum	0·06

    Show, as a simple whole number ratio, the relative amounts of cholesterol in liver plasma membrane, mitochondrial membrane and endoplasmic reticulum.   1

    (5)

Marks

6. Binding of specific proteins to DNA is important in the control of gene expression.

   (a) Describe the effect of repressor protein binding to DNA in the *lac* operon.

1

   (b) Binding to DNA of the *engrailed* protein of the fruit fly *Drosophila melanogaster* is required during development of the fruit fly embryo.

The DNA-binding region of the engrailed protein consists of a stretch of sixty amino acids that contain two α-helices connected by a short extended chain of amino acids as shown in Figure 1.

Figure 1

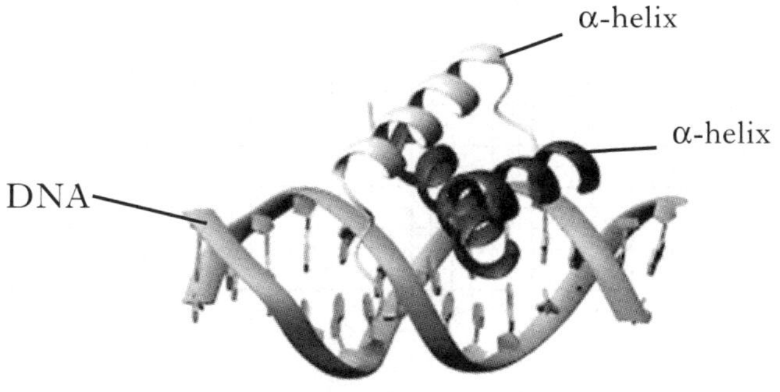

     (i) What level of protein structure is an α-helix?

1

     (ii) The side chains of the amino acids within the α-helix regions interact directly with DNA. Figure 2 shows the amino acids in a short stretch of one of the α-helix regions of the engrailed protein.

Figure 2

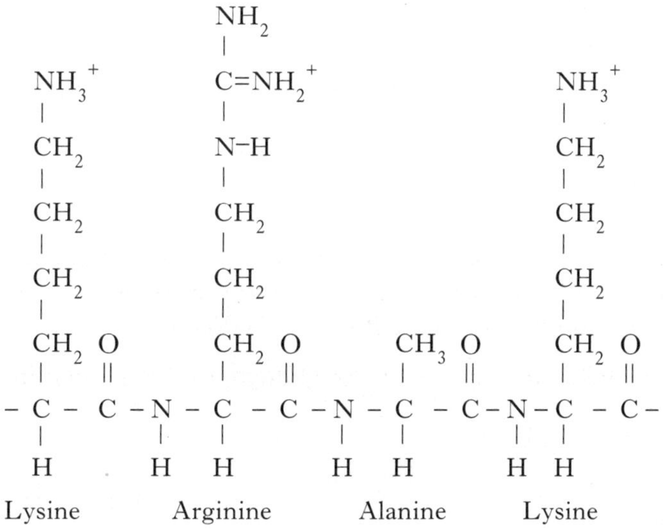

          Name the class of amino acid to which lysine belongs.

1

     (iii) The binding region of the engrailed protein contains a high proportion of lysine residues. Suggest how the presence of these amino acids would assist in the binding of the engrailed protein to DNA.

1

(4)

[Turn over

Marks

7. Enzyme kinetics is the study of the rate of enzyme-catalysed reactions.

The graph below shows the rates of the reaction for the enzyme penicillinase over a range of substrate concentrations. The substrate is penicillin.

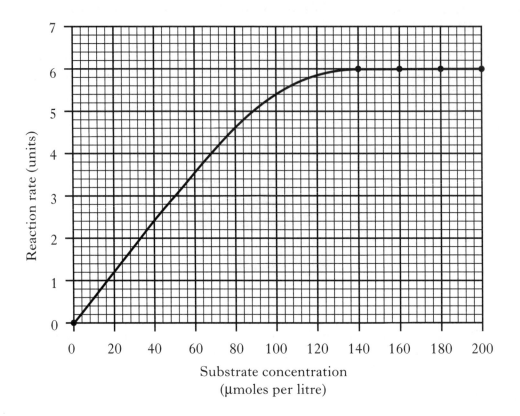

The Michaelis constant (K_m) of an enzyme is the substrate concentration at which the reaction rate is half its maximum rate.

(a) Calculate the K_m of penicillinase assuming the graph shows that the maximum rate has been reached.   **1**

(b) Explain why the K_m of an enzyme increases when a competitive inhibitor is present.   **1**

(c) The turnover number of an enzyme is the number of substrate molecules converted into product by an enzyme in one second when an enzyme is working at its maximum rate. The turnover number for penicillinase is 2000 per second.

Calculate the time taken to catalyse the breakdown of one penicillin molecule.   **1**

      (3)

8. Answer **either** A **or** B.

   A. Describe the structure of the monosaccharide glucose. Discuss the structures and functions of the main polysaccharides made using glucose as a monomer.   **(15)**

   OR

   B. Give an account of the processes involved in the polymerase chain reaction (PCR) and DNA profiling.   **(15)**

[END OF SECTION B]

SECTION C

Candidates should attempt questions on <u>one</u> unit, <u>either</u> Biotechnology <u>or</u> Animal Behaviour <u>or</u> Physiology, Health and Exercise.

The questions on Animal Behaviour can be found on pages 20–23.

The questions on Physiology, Health and Exercise can be found on pages 24–26.

All answers must be written clearly and legibly in ink.

Labelled diagrams may be used where appropriate.

BIOTECHNOLOGY   *Marks*

1. The Figure below shows the final stage in a test that confirms a blood sample contains antibodies against *Herpes simplex* virus (HSV). HSV antigen is attached to the plastic well and any unbound areas are coated with non-reactive material.

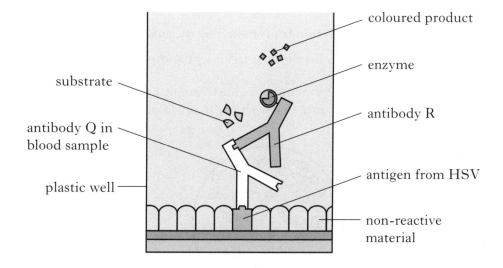

(a) (i)   Identify the technique represented in the diagram.   1

     (ii) Explain why the test represented would not reveal if the person had been infected with chickenpox virus.   1

     (iii) Use information in the Figure to explain why inadequate rinsing just before the stage shown might result in a *false* positive result.   2

(b) Antibody R was produced commercially in a fermenter from hybridoma cells.

     What **two** cell types are combined to make hybridoma cells?   1

                                                                                   (5)

[Turn over

BIOTECHNOLOGY (continued)   *Marks*

2.   The flow chart shows steps involved in the manufacture of yoghurt.

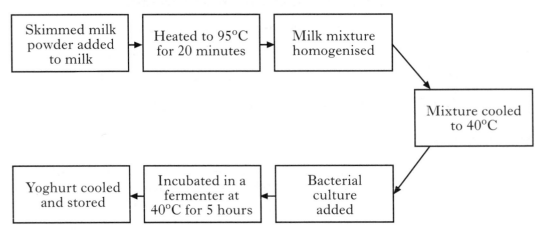

(a) The milk mixture is heated to 95 °C for 20 minutes to remove dissolved oxygen.

    (i)   What chemical conversion is promoted by the anaerobic conditions?   1

    (ii) Give a further reason for the heat treatment at this stage.   1

(b) During incubation in the fermenter, yoghurt samples were removed and examined under a microscope. The figure below shows the field of view.

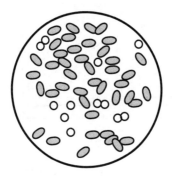

    (i)   What can be observed?   1

    (ii) Account for the observation.   1

      (4)

3.   Describe the scaling up process required to produce a suitable volume of pure bacterial culture for an industrial fermenter.   **(5)**

BIOTECHNOLOGY (continued)   *Marks*

4.   Silage is used for winter feeding of farm animals and is commonly made by wrapping baled grass in polythene. Ensilage of plants in this way preserves the nutritional quality by limiting protein breakdown.

(a) Apart from wrapping bales, give **one** other method of making silage.   1

(b) Explain how changes that take place within the wrapped bale help to preserve the silage.   2

(c) Name a bacterial species that would be added before the baled grass is wrapped.   1

(d) In a study involving ensilage of harvested lupin plants, researchers evaluated the effect of adding bacteria to the bales. The graphs below show the data obtained. Error bars show variation between replicates.

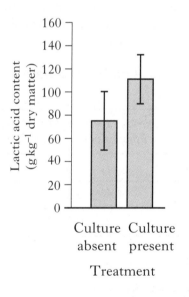

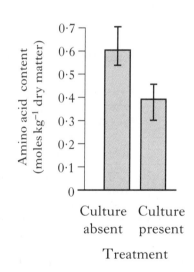

(i) What evidence supports the conclusion that the treatment with bacteria preserves the nutritional quality of the plant material?   1

(ii) Draw **one** other conclusion about the effect of adding bacteria to fermenting silage.   1

(6)

(20)

[End of *Biotechnology* questions. *Animal Behaviour* questions start on Page 20]

SECTION C (continued)   *Marks*

ANIMAL BEHAVIOUR

1. The eastern spinebill (*Acanthorhyncus tenuirostris*) is a small bird from eastern Australia. One of its major foods is nectar from the mountain correa (*Correa lawrenciana*).

    Figure 1: Eastern spinebill feeding on mountain correa

    (*a*) In a study of eastern spinebill foraging, flowers of mountain correa were assigned to different developmental stages (Floral stages 1–5). Some characteristics of each stage are shown in the Table. Figure 2 shows the abundance of floral stages available and the foraging choices made by eastern spinebills feeding on the flowers.

    Table: Characteristics of mountain correa flowers at different stages

Floral stage	*Age (days)*	*Pollen production*	*Volume of nectar produced per flower over 24 hours (μ l)*
1	1–2	Pollen present, not released	1·0
2	3–7	Pollen released	3·1
3	8–9	Little, if any, pollen present	0·5
4	10–13	No pollen	0
5	14+	No pollen	0

ANIMAL BEHAVIOUR (continued)   *Marks*

1. **(*a*) (continued)**

Figure 2: Proportions of floral stages available and foraging choices made by eastern spinebills

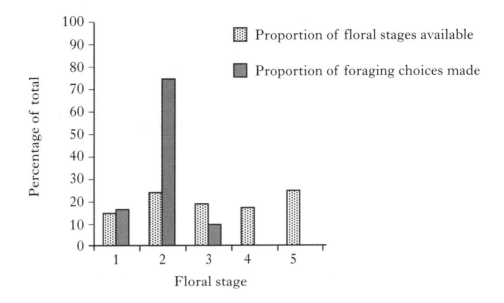

    (i)   Use the data in Figure 2 to compare floral stages 2 and 5.   2

    (ii) What is meant by the term optimal foraging?   1

    (iii) How does the information provided in both the Table and Figure 2 demonstrate optimal foraging in spinebills?   1

(*b*) The eastern spinebill's nest is a small cup made mainly from twigs, grass, bark, feathers and spider webs. Only the female builds the nest and incubates the eggs but both parents feed the young after they have hatched.

    (i)   This information suggests greater investment by the female parent. Describe another way in which female investment is likely to be greater than that of the male parent.   1

    (ii) Explain how nest building in the eastern spinebill provides an example of an extended phenotype.   1

(*c*) The eastern spinebill does not show any marked sexual dimorphism.

Figure 1 shows a male bird. What would the female bird look like in comparison?   1

                                                                                        (7)

[Turn over

ANIMAL BEHAVIOUR (continued)   *Marks*

2. The silk produced by female spiders often contains chemical deposits that provide males with important information about species identity, age, sex and the reproductive status of a female.

Female wolf spiders *Schizocosa ocreata* rarely mate with more than one male and, once they have mated, are more likely to cannibalise (eat) males. Males, on the other hand, will often try to mate with more than one female.

The male has a courtship behaviour called a "jerky tap" that elicits a reaction from the female. The Table below shows measurements of the time taken for males to produce the jerky tap response after exposure to samples of silks from different origins.

Origin of silk	*Time until jerky tap response* (s)
Subadult female	155
Unmated adult female	22
Mated adult female	105

(a) (i)   State **one** conclusion that can be drawn from these results.   1

(ii)   What name is given to the delay between stimulus and response?   1

(iii)   Name another aspect of this jerky tap response that could be observed and used for comparison.   1

(iv)   Suggest **one** disadvantage of laboratory-based research into animal behaviour.   1

(b) The "selfish gene" concept maintains that individual organisms should behave so as to maximise the survival of copies of their genes.

Give **one** reason why the genes responsible for a male spider's response to silk in the selection of a mate can be described as "selfish".   1

(5)

ANIMAL BEHAVIOUR (continued)   *Marks*

3. Cheetahs (*Acinonyx jubatus*) in the Serengeti National Park in Tanzania kill more male Thomson's gazelles (*Gazella thomsoni*) than would be expected from the sex ratio of the local gazelle population.

The Table below shows data obtained by observing groups of Thomson's gazelles.

	males	*females*
Proportion on periphery of group (%)	75	53
Nearest neighbour distance (m)	9·3	4·6
Proportion of time spent scanning with head up (%)	8·4	11·4
Proportion in population (%)	30	70
Proportion hunted (%)	63	37

(a) Suggest why male Thomson's gazelles are more likely than females to be selected as prey by hunting cheetahs.   1

(b) Calculate the number of males on the periphery of a group of 80 gazelles.   1

(c) What name is given to the scanning behaviour?   1

                                                                                  (3)

4. Describe how appeasement and ritualised displays in agonistic interactions can benefit all members of social groups. Illustrate your answer by reference to named species.   (5)

                                                                                  (20)

[End of *Animal Behaviour* questions. *Physiology, Health and Exercise* questions start on Page 24]

[Turn over

SECTION C (continued)   *Marks*

PHYSIOLOGY, HEALTH AND EXERCISE

1. (*a*) The graph shows obesity data for England in 1993 and 2002.

Individuals were described as obese if they had a body mass index (BMI) of 30 or greater.

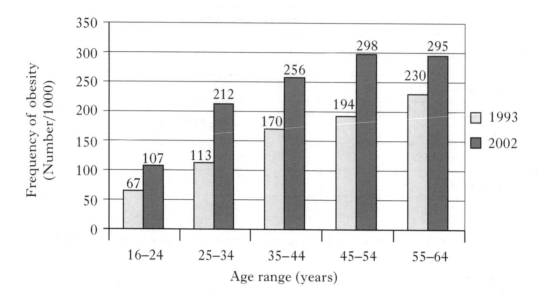

	(i)	What two measurements are needed to calculate BMI?	1
	(ii)	Obesity has increased in all age ranges over the ten year period.	
		Which age range has shown the biggest percentage increase?	1
	(iii)	Give **one** other general trend shown by the data.	1
	(iv)	Name **one** medical condition for which obesity is a risk factor.	1

(*b*) (i)   Bioelectrical impedance analysis (BIA) is a method used to determine percentage body fat. Outline the principle on which this method is based.   **2**

(ii) Give **one** limitation of BIA.   **1**

                                                                  (7)

2. Discuss the changes that take place in the cardiovascular system during a short period of strenuous exercise.

                                                                  (4)

PHYSIOLOGY, HEALTH AND EXERCISE (continued)   *Marks*

3. The Bruce protocol is a method used in maximal exercise testing to determine fitness. A subject wearing an oxygen-monitoring mask is supervised running on a treadmill while the gradient (slope) and speed of the treadmill are both increased in a standard way. When the subject is exhausted, the time is noted.

   The Table below shows some results from a study using this method. The values have been selected for four young healthy males who each took the same time to reach exhaustion.

Time (min)	Body mass (kg)	Maximum oxygen uptake ($l \, min^{-1}$)	Fitness ($ml \, kg^{-1} \, min^{-1}$)
10·5	70	2·53	36·2
10·5	75	2·72	36·2
10·5	85	3·08	36·2
10·5	90	3·26	36·2

   (a) Explain why measuring oxygen uptake is a valid way to assess fitness.   2

   (b) Explain why the "Fitness" measurement is independent of body mass.   1

   (c) Calculate the maximum oxygen uptake of an 80 kg male who took 10·5 minutes to reach exhaustion in the same test.   1

   (d) Give an example of a situation where the individual given a treadmill test would **not** be stressed to exhaustion.   1

        (5)

[Turn over for Question 4 on *Page twenty-six*

PHYSIOLOGY, HEALTH AND EXERCISE (continued)   *Marks*

4.   In an investigation into energy expenditure measured by direct calorimetry, subjects at
     rest were given a solution of either glucose or sucrose (a disaccharide). The results are
     shown in the graph below.

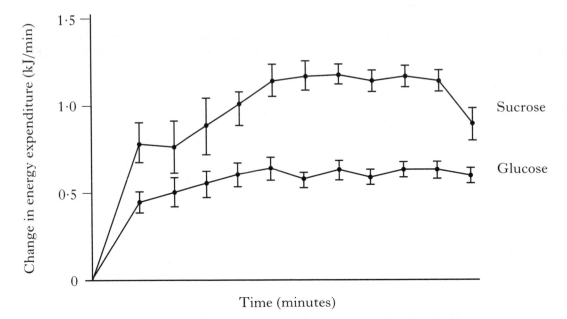

(*a*) Which component of total energy expenditure is being investigated in this study?   **1**

(*b*) What evidence is there that diet affects energy expenditure?   **1**

(*c*) What do the error bars in the graph indicate about the results presented?   **1**

(*d*) The results in this investigation were obtained by direct calorimetry.

       Give **one** way in which indirect calorimetry differs from direct calorimetry.   **1**

                                                                                              (4)

                                                                                             (20)

[*END OF QUESTION PAPER*]

[BLANK PAGE]

X007/701

| NATIONAL QUALIFICATIONS 2010 | THURSDAY, 27 MAY 1.00 PM – 3.30 PM | BIOLOGY ADVANCED HIGHER |

SECTION A—Questions 1–25 (25 marks)

Instructions for completion of Section A are given on *Page two*.

SECTIONS B AND C

The answer to each question should be written in ink in the answer book provided. Any additional paper (if used) should be placed inside the front cover of the answer book.

Rough work should be scored through.

Section B (55 marks)

All questions should be attempted. Candidates should note that Question 8 contains a choice.

Question 1 is on Pages 10, 11 and 12. Questions 2 and 3 are on Page 13. Pages 12 and 13 are fold-out pages.

Section C (20 marks)

Candidates should attempt the questions in **one** unit, **either** Biotechnology **or** Animal Behaviour **or** Physiology, Health and Exercise.

Read carefully

1   Check that the answer sheet provided is for **Biology Advanced Higher (Section A)**.

2   For this section of the examination you must use an **HB pencil** and, where necessary, an eraser.

3   Check that the answer sheet you have been given has **your name**, **date of birth**, **SCN** (Scottish Candidate Number) and **Centre Name** printed on it.

    Do not change any of these details.

4   If any of this information is wrong, tell the Invigilator immediately.

5   If this information is correct, **print** your name and seat number in the boxes provided.

6   The answer to each question is **either** A, B, C or D. Decide what your answer is, then, using your pencil, put a horizontal line in the space provided (see sample question below).

7   There is **only one correct** answer to each question.

8   Any rough working should be done on the question paper or the rough working sheet, **not** on your answer sheet.

9   At the end of the examination, put the **answer sheet for Section A inside the front cover of the answer book**.

Sample Question

Which of the following molecules contains six carbon atoms?

A   Glucose

B   Pyruvic acid

C   Ribulose bisphosphate

D   Acetyl coenzyme A

The correct answer is **A**—Glucose. The answer **A** has been clearly marked in **pencil** with a horizontal line (see below).

Changing an answer

If you decide to change your answer, carefully erase your first answer and using your pencil, fill in the answer you want. The answer below has been changed to **D**.

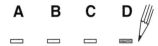

SECTION A

All questions in this section should be attempted.

Answers should be given on the separate answer sheet provided.

1. The following diagram represents a bacterial cell.

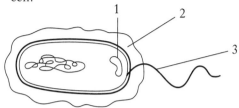

Which **one** of the following correctly identifies the structures labelled 1, 2 and 3?

	1	2	3
A	plasmid	flagellum	capsule
B	flagellum	capsule	plasmid
C	plasmid	capsule	flagellum
D	capsule	plasmid	flagellum

2. Plasmodesmata are structures that link

  A   cell walls in adjacent prokaryotic cells

  B   cell walls in adjacent eukaryotic cells

  C   cell cytoplasm in adjacent prokaryotic cells

  D   cell cytoplasm in adjacent eukaryotic cells.

3. The key below shows features of biological molecules.

1   Soluble in water.......................go to 2
     Insoluble in water...................go to 3

2   Extracellular...........................A
     Hydrolytic..............................B

3   Storage...................................C
     Structural...............................D

Which letter could be both triglycerides and glycogen?

4. The diagram below shows the changes in cell mass and DNA mass during two cell cycles.

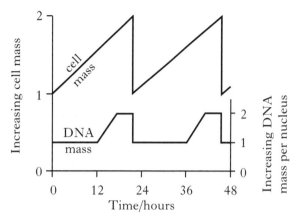

What valid conclusion could be made from the graph?

During the cell cycle

  A   interphase is the longest phase

  B   mitosis is divided into four phases

  C   replication takes place between 0 and 12 hours

  D   cytokinesis takes place at 12 and 36 hours.

5. Which line in the table below correctly describes adenine and thymine and the bonding between them in a DNA molecule?

	Adenine	Number of hydrogen bonds	Thymine
A	purine	two	pyrimidine
B	pyrimidine	three	purine
C	pyrimidine	two	purine
D	purine	three	pyrimidine

[Turn over

6. One cause of cystic fibrosis is a mutation in the CFTR gene which codes for 1480 amino acids. The most common mutation results in the deletion of one amino acid.

Which line in the table below shows correctly the number of nucleotides in the mutated gene and the number of amino acids in the protein that is synthesised?

	Number of nucleotides encoding the mutated gene	Number of amino acids in the protein synthesised
A	4431	1477
B	4439	1480
C	4437	1479
D	4439	1479

7. The diagram below shows the distribution of protein molecules in a cell membrane.

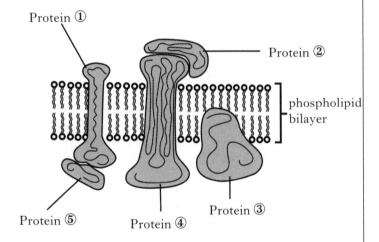

Protein ①

Protein ②

phospholipid bilayer

Protein ③

Protein ④

Protein ⑤

Which line in the table below correctly identifies a peripheral and an integral membrane protein?

	Peripheral membrane protein	Integral membrane protein
A	1	5
B	2	1
C	3	4
D	5	2

8. The diagram below shows a metabolic pathway that is controlled by end product inhibition.

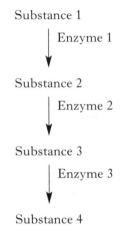

Substance 1

Enzyme 1

Substance 2

Enzyme 2

Substance 3

Enzyme 3

Substance 4

For Substance 4 to bring about end product inhibition, which of the following will it interact with?

A   Substance 1

B   Substance 3

C   Enzyme 1

D   Enzyme 3

9. Covalent modification of enzymes is used to control their activity.

Which of the following processes involves the covalent modification of an enzyme?

A   The conversion of trypsinogen into trypsin.

B   The end-product inhibition of phosphatase.

C   The allosteric inhibition of glycogen phosphorylase.

D   The conversion of sucrose into glucose and fructose.

10. Which of the following acts as a hydrophobic extracellular signalling molecule?

A   Insulin

B   Testosterone

C   Acetylcholine

D   Cholesterol

11. The following stages are involved in amplifying DNA fragments using the polymerase chain reaction (PCR).

V   DNA denatures

W   Primers bind

X   Complementary DNA strands replicated

Y   Temperature changed to about 75 °C

Z   Temperature changed to about 55 °C

After heating the fragments to about 95 °C, which of the following sequences occurs?

A   X, Z, V, Y, W

B   Z, V, W, Y, X

C   V, X, Z, W, Y

D   V, Z, W, Y, X

12. A piece of DNA 20 kilobase pairs (kbp) long was digested using different restriction enzymes. BamHI, EcoRI and PstI. The results are shown in the table below.

	Restriction enzyme used			
	PstI	BamHI	EcoRI	BamHI PstI
Lengths of DNA fragments (kbp)	15	17	12	12
	5	3	8	5
				3

Which of the following set of fragments would result if all three enzymes were used together?

A   7, 5, 3, 2

B   8, 7, 3, 2

C   12, 3, 2

D   9, 5, 3, 3

13. A certain restriction enzyme will only cut a DNA strand between two Cs when the base sequence CCGG is present.

Two homologous segments of DNA that carry different alleles of a gene are shown below. Allele 2 has a single base-pair difference.

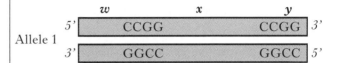

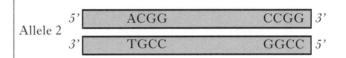

Allele 1 was treated with the restriction enzyme and the fragments *w*, *x* and *y* were obtained and separated by gel electrophoresis. The resulting band pattern for allele 1 is shown below.

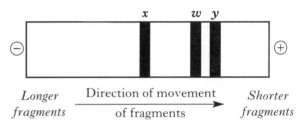

Longer fragments   Direction of movement of fragments   *Shorter fragments*

Which of the following band patterns would result when the procedure was repeated using allele 2?

A

B

C

D

14. Several species of bacteria have been found deep under the Pacific Ocean, where hot water escapes from the sea bed.

In these marine ecosystems, the bacteria can use hydrogen sulphide as an energy source to fix carbon dioxide into organic molecules. When the bacteria break down, organic material is released, which filter feeders consume.

This information indicates that the bacteria are

A   heterotrophs

B   detritivores

C   autotrophs

D   decomposers.

15. Three pyramids of numbers are shown below.

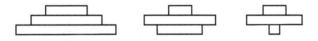

Which of the following food chains **cannot** be represented by any of these pyramids?

A   Oak tree → leaf miner → tree warbler

B   Algae → pond snail → nematode parasite

C   Grass plants → rabbit → stoat

D   Phytoplankton → zooplankton → herring

16. The table below shows data obtained from an investigation into the mass and population density of some organisms in a heathland food web.

Which line in the table shows correctly the species with the highest biomass per square metre?

	Species	Mean mass of organism (g)	Population density (numbers m^{-2})
A	cricket	0·10	4
B	ladybird	0·03	20
C	aphid	0·002	5420
D	weevil	0·005	3250

17. The diagram below shows the nitrogen cycle.

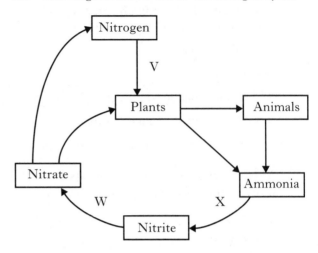

Which line in the table below correctly identifies the micro-organisms involved at the stages shown?

	V	W	X
A	Nitrosomonas	Nitrobacter	Rhizobium
B	Rhizobium	Nitrobacter	Nitrosomonas
C	Nitrobacter	Rhizobium	Nitrosomonas
D	Rhizobium	Nitrosomonas	Nitrobacter

18. The table below shows four examples of interactions between species.

Which column in the table shows correctly the benefits (+) or costs (−) which result from each interaction?

Interaction	A	B	C	D
Sheep grazing in a field of grass	+/−	+/−	+/+	+/−
Owls and foxes hunting for the same food	+/−	−/−	−/−	+/−
Corals acting as hosts for zooxanthellae	+/−	+/+	+/+	+/+
"Cleaner fish" feeding on parasites which they remove from other fish	+/+	+/+	+/−	+/+

19. Anolis lizards are found on Caribbean islands. They feed on prey of various sizes.

Histogram 1 shows the range of prey length eaten by *Anolis marmoratus* on the island of Jarabacoa, where there are five other Anolis species.

Histogram 2 shows the range of prey length eaten by *Anolis marmoratus* on the island of Marie Galante, where it is the only Anolis species.

Histogram 1: Jarabacoa Island

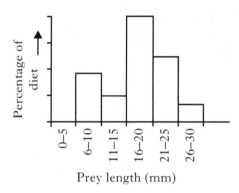

Prey length (mm)

Histogram 2: Marie Galante Island

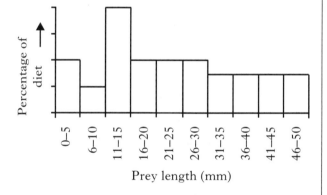

Prey length (mm)

Which of the following statements could explain the different range of prey sizes eaten by *Anolis marmoratus* on the two islands?

A Larger numbers of prey are found on Marie Galante.

B *Anolis marmoratus* occupies its fundamental niche on Jarabacoa.

C *Anolis marmoratus* occupies its realised niche on Marie Galante.

D Resource partitioning takes place on Jarabacoa.

20. Each statement below applies to either conformation or regulation.

1 A wide range of habitats can be occupied
2 A restricted range of habitats can be occupied
3 There is a high energy cost
4 There is no energy cost

Which statements apply to conformation?

A 1 and 4

B 1 and 3

C 2 and 4

D 2 and 3

21. The figure below shows the general relationships between the internal environment and variation in the external environment of four animals.

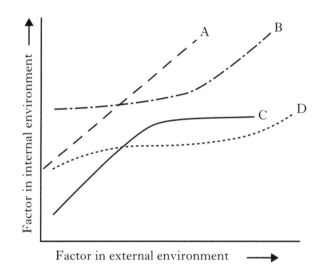

Which animal is the most effective regulator?

[Turn over

22. Eutrophication can result from agricultural activity.

Which of the following defines eutrophication?

A   Algal bloom

B   Increased BOD

C   Loss of diversity

D   Nutrient enrichment

23. The figure below represents part of an aquatic food web.

P1 and P2 are producers.

C1, C2 and C3 are consumers.

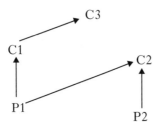

Analysis of a persistent organic pesticide in this ecosystem produced the following results:

Result 1 P1 has a higher concentration of the pesticide in its tissues than is present in the surrounding water.

Result 2 C2 converts the pesticide into a more toxic chemical in its tissues.

Result 3 The concentration of the pesticide in P1 is lower than that in C1 which, in turn, is lower than that in C3.

Which row in the table shows the processes responsible for Results 1, 2 and 3?

	Result 1	Result 2	Result 3
A	bioaccumulation	biotransformation	biomagnification
B	biomagnification	biotransformation	bioaccumulation
C	biotransformation	bioaccumulation	biomagnification
D	bioaccumulation	biomagnification	biotransformation

24. The insecticide DDT is metabolised in birds to DDE. The level of DDE in eggs affects the shell thickness. Premature egg breakage begins when mean shell thickness is 80% of normal.

The graph below shows how DDE content of the diet affects DDE content of eggs and mean shell thickness.

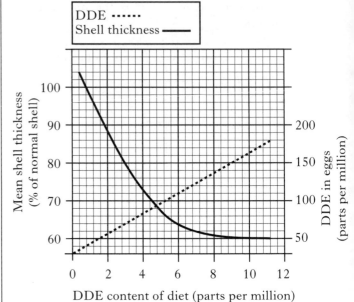

What would be the minimum DDE concentration to cause the start of premature egg breakage?

A   150 parts per million

B   70 parts per million

C   9 parts per million

D   3 parts per million

25. Succession which follows the clearing of long-established farm land is described as

A   allogenic

B   secondary

C   degradative

D   primary.

[END OF SECTION A]

Candidates are reminded that the answer sheet MUST be returned INSIDE the front cover of the answer book.

[Turn over for Section B on *Page ten*

SECTION B

All questions in this section should be attempted.
All answers must be written clearly and legibly in ink.

1. Proto-oncogenes code for proteins that stimulate cell division. When these genes mutate they can become oncogenes and cause the excessive cell proliferation associated with tumour formation.

   A recent study investigated a chromosome mutation discovered in lung tumour cells. The mutation is a rearrangement of two genes on the same chromosome, resulting in the fusion of the two genes (Figure 1). The genes are *EML4*, which is involved in microtubule formation, and *ALK* which codes for a kinase enzyme. As a result of this fusion, a new protein is formed that has uncontrolled kinase activity inside the cell.

   Figure 1 : Formation of the EML4–ALK fusion gene

   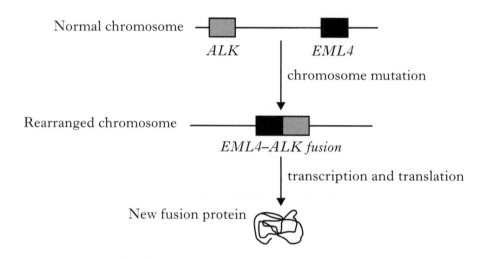

   Part of the study aimed to find out if normal cells are transformed to divide abnormally after treatment with the fusion gene. In the treatments, genes were introduced into normal cells that were then cultured in a flat dish containing a suitable growth medium. Abnormal cell division is indicated by the transformed cells stacking up in multiple layers called *foci*.

   The four gene treatments used are shown in the Table. In the fourth, the *EML4–ALK* fusion gene was modified so that the kinase produced was inactive.

   Table: Results of cell transformation study

Gene treatment	*Formation of foci*
ALK alone	no
EML4 alone	no
EML4–ALK fusion	yes
EML4–ALK fusion modified	no

Question 1 (continued)

The study also investigated the possibility that kinase inhibitors could be used to treat tumours arising from the fusion gene. The growth of normal and transformed cells suspended in liquid culture was monitored in the presence and absence of a kinase inhibitor. The results are shown in Figures 2 and 3.

Figure 2: Effect of kinase inhibitor on normal cells

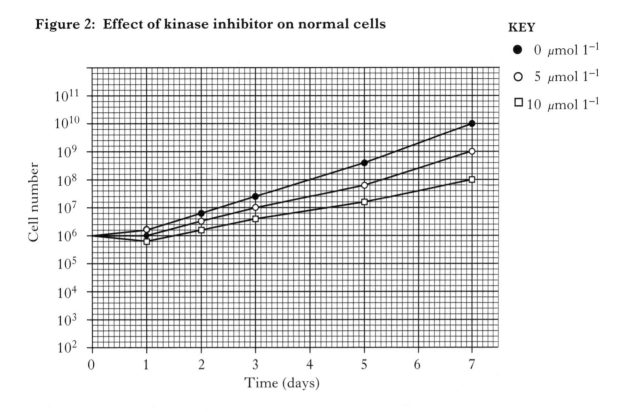

Figure 3: Effect of kinase inhibitor on cells transformed with *EML4–ALK*

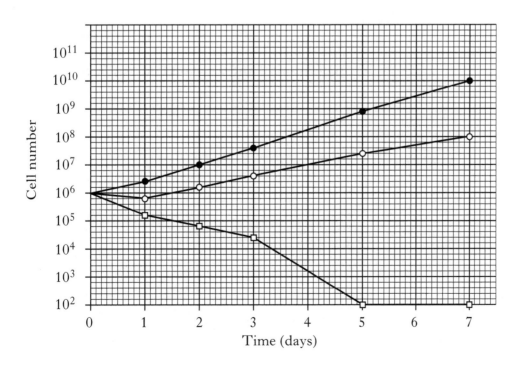

[Question 1 continues on *Page twelve*

Marks

Question 1 (continued)

(a) State why abnormal proliferation of cells can occur when only one copy of an oncogene is present.   1

(b) (i) In the study, abnormal cells formed *foci*. Describe the appearance of a culture of normal animal cells that have stopped dividing.   2

     (ii) Give a reason for adding fetal bovine serum to cell culture medium.   1

(c) Refer to the table. Explain how the results show that:

     (i) the chromosome rearrangement created an oncogene;   1

     (ii) the effect of the fusion gene is uncontrolled kinase activity.   2

(d) Use data from Figures 2 and 3 to show that normal and transformed cells are capable of dividing at the same rate.   1

(e) Refer to Figures 2 and 3.

     (i) Show that the $10\,\mu\text{mol}\,l^{-1}$ inhibitor concentration reduces the proliferation of normal cells by a factor of 100.   1

     (ii) What general trend is observed for the growth of cells when the concentration of inhibitor is varied?   1

     (iii) Use data from the $5\,\mu\text{mol}\,l^{-1}$ concentration to show that transformed cells are more sensitive to the presence of the inhibitor than normal cells.   1

(f) The study aimed to find a possible therapy for tumours caused by fusion mutation. How do the results suggest that $10\,\mu\text{mol}\,l^{-1}$ of inhibitor would be the most useful to test on patients?   2

(13)

[Questions 2 and 3 are on fold-out *Page thirteen*

Marks

2. The cell wall of the prokaryote *E. coli* is made of a substance that consists of polysaccharide chains cross-linked by short chains of amino acids. The polysaccharide is made up of two kinds of sugar: N-acetylmuramate (NAM) and N-acetylglucosamine (NAG). NAM and NAG alternate along the chain and differ from glucose only at the C2 and C3 positions.

(a) Name the substance that makes up the cell wall of *E. coli*. **1**

(b) NAM and NAG are joined by a glycosidic bond.

   (i) Explain why this bond is described as $\beta(1-4)$. **2**

   (ii) The enzyme lysozyme damages bacterial cell walls by breaking the bonds between NAM and NAG.

     What type of reaction is catalysed by lysozyme? **1**

               (4)

3. Describe the transport of sodium and potassium ions across the plasma membrane. **(5)**

Marks

4. Transgenic plants can be produced using genetically engineered plasmids. The plasmids are obtained from bacteria that naturally infect plant cells. A modified plasmid is shown in the diagram below.

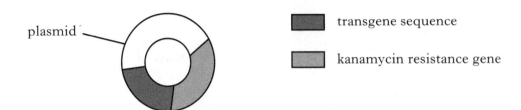

Bacterial cells containing the modified plasmid are incubated with plant cells. The plant cells are then cultured in a growth medium containing kanamycin.

(a) What is meant by the term transgenic?   1

(b) Name the bacterial source of these plasmids.   1

(c) Explain the role of kanamycin in the production of transgenic plants.   2

(d) Give **one** example of the use of this transgenic technology.   1

    (5)

Marks

5.   Bramble plants (*Rubus fruticosus*) are pollinated by a variety of nectar-feeding insects, such as the meadow brown butterfly (*Maniola jurtina*). Bramble flowers are one of many nectar sources for this species.

A study focused on competitive interactions occurring between meadow browns and other insects at bramble flowers. The average time a meadow brown spent feeding when not disturbed by another insect is shown in the bar graph at A. The other bars show its feeding duration when another insect was also present.

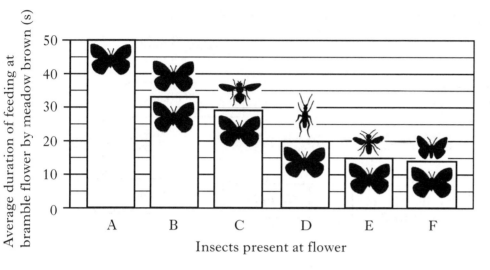

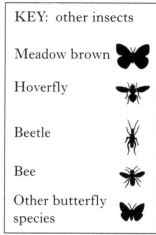

(a)   Explain why intraspecific competition is expected to be more intense than interspecific.   1

(b)   (i)   Which bar represents intraspecific competition?   1

(ii)   Give **one** general conclusion that can be drawn from these data about competitive interactions at bramble flowers?   1

(c)   From the information provided, state why the relationship between meadow brown butterflies and bramble plants is not an example of mutualism even though there are benefits to both species.   1

                                                                                                                                          (4)

[Turn over

Marks

6. A laboratory experiment investigated the decomposition of beech (*Fagus sylvatica*) leaves following four different treatments. Fresh leaves were either left to form leaf litter or were fed to herbivorous caterpillars to form caterpillar faeces. The mass losses of these samples were then compared with and without the detritivore activity of woodlice.

Figure 1 shows the four treatments carried out. Figure 2 shows the mass loss of the four samples produced over a twelve week period.

Figure 1

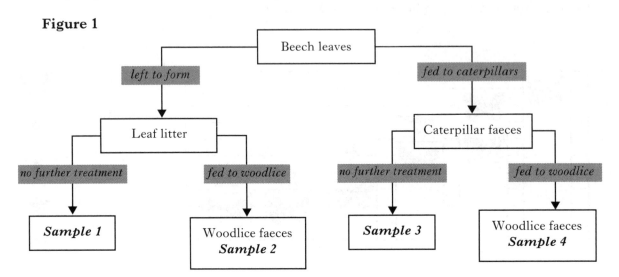

Figure 2

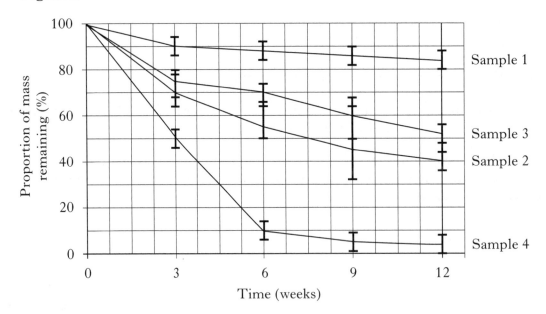

(a) How do the diets of herbivores and detritivores differ?   1

(b) Refer to Figure 2.

   (i) What evidence is there that leaves passing through the guts of two invertebrates decompose more than seven times faster than leaves allowed to form leaf litter?   2

   (ii) Use the error bars to comment on the results obtained for Samples 2 and 3 after 9 weeks.   1

   (iii) Explain why Sample 4 shows the most rapid loss in mass.   1

                                                                          (5)

Marks

7. (*a*) Explain how the level of pollution in an ecosystem could be monitored using changes in population.   **2**

   (*b*) The widespread use of the drug diclofenac to treat cattle in the Indian subcontinent has led to a rapid decline in the populations of various species of vulture. For example, in 2008 the population of the oriental white-backed vulture (*Gyps bengalensis*) was estimated to be 11 000, a decline of 99·9% since 1992. Although diclofenac has low persistence and generally it has low toxicity, vultures are now known to be unusually susceptible to it.

   (i) What was the population of *Gyps bengalensis* in 1992?   **1**

   (ii) Explain why the effect on the vulture populations is greatest when they feed on carcasses of cattle treated with diclofenac shortly before their death.   **1**

   (4)

8. Answer **either** A **or** B.

   A. Increasing the rate of food production to meet global demands is challenging.

   Discuss the management of the environment for intensive food production with reference to:

   (i) control of species that reduce yield;   **7**

   (ii) monoculture.   **8**

   OR   **(15)**

   B. Plants are increasingly being cultivated for biomass as an alternative source of energy to fossil fuels.

   Discuss:

   (i) energy fixation and primary productivity;   **5**

   (ii) fossil fuels and air pollution.   **10**

   (15)

[*END OF SECTION B*]

[Turn over

SECTION C

Candidates should attempt questions on <u>one</u> unit, <u>either</u> Biotechnology <u>or</u> Animal Behaviour <u>or</u> Physiology, Health and Exercise.

The questions on Biotechnology can be found on pages 18–20.

The questions on Animal Behaviour can be found on pages 22–25.

The questions on Physiology, Health and Exercise can be found on pages 26–28.

All answers must be written clearly and legibly in ink.

Labelled diagrams may be used where appropriate.

BIOTECHNOLOGY   *Marks*

1. (*a*) Legumes interact with bacteria to produce nitrogenase.

      (i) Name the genes responsible for the synthesis of nitrogenase.   **1**

      (ii) What chemical transformation is catalysed by nitrogenase?   **1**

  (*b*) Sea buckthorn (*Hippophae rhamnoides*) produces fruit that has potential as a food and medicinal crop. Although not a legume, it has a symbiotic relationship in which nitrogenase is synthesised and the plant receives nitrate. When it is grown as a crop, nitrogen fertiliser is required to produce a high yield of fruit.

    The Figure below shows the effect of daily nitrate fertiliser application on nitrogenase activity in sea buckthorn root nodules over a period of 30 days.

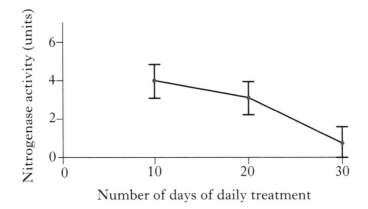

    Describe the effect of nitrate application on nitrogenase activity.   **2**

                                                        (4)

2. Describe how B lymphocytes respond to foreign antigens. How is this response applied in the preparation of polyclonal sera?   **(5)**

Marks

BIOTECHNOLOGY (continued)

3. Bananas are important as a crop but the planting material required is often in short supply. Scientists have developed a micropropagation system as shown in the flow chart.

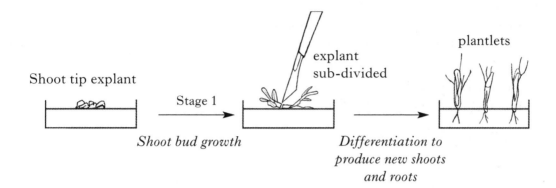

Shoot tip explant

Stage 1

Shoot bud growth

explant sub-divided

Differentiation to produce new shoots and roots

plantlets

(a) Suggest **two** advantages of propagating banana plants using this system.   2

(b) State **two** environmental conditions that need to be controlled during the development of the plantlets.   1

(c) The effect of plant growth regulators on shoot regeneration in Stage 1 is shown in the table.

MS medium + growth regulators (mg l^{-1})		Explants differentiated (%)	Mean number of shoots per explant
+ IAA	+Kinetin		
0	0	29	2
0·1	1·0	42	4
0·1	2·5	46	8
0·1	5·0	58	8
0·5	1·0	54	12
0·5	2·5	66	20
0·5	5·0	58	28

   (i) What evidence is there that growth regulators benefit propagation?   1

   (ii) Use the data to explain why a growth regulator combination of 0·5 mg l^{-1} IAA + 5·0 mg l^{-1} kinetin is recommended.   1

   (5)

[Turn over

Marks

BIOTECHNOLOGY (continued)

4. (a) Probiotics are produced by the dairy industry to provide health benefits beyond basic nutrition.

Describe **one** such health benefit.

1

(b) A company that produces a probiotic yoghurt drink claims there is a measurable beneficial effect when there are at least 10^{10} viable cells of *Lactobacillus casei* per $100 \, cm^3$ carton.

The Figure shows how a biotechnologist used dilution plating to check the number of bacteria in a $100 \, cm^3$ carton of the drink.

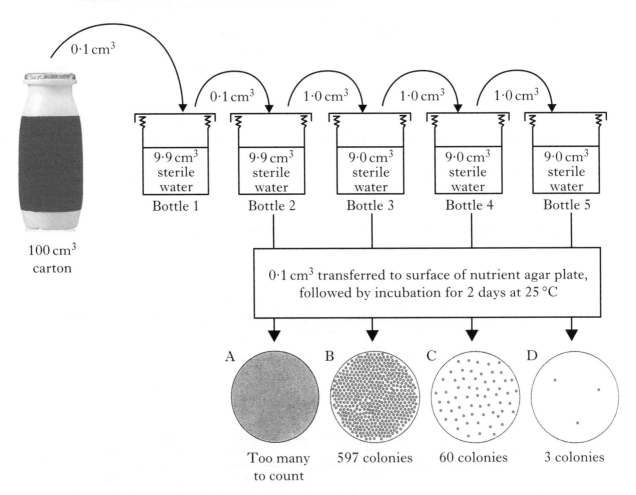

(i) Explain why dilution plating was used to check the number of bacteria rather than a direct count under the microscope.

1

(ii) Explain why sterile water was used throughout the dilution steps.

1

(iii) Explain why plate C would be selected to check the number of bacteria in the drink.

1

(iv) Use the information to show that more than 10^{10} viable cells were present in the carton.

2

(6)

(20)

[End of *Biotechnology* questions. *Animal Behaviour* questions start on Page 22]

[BLANK PAGE]

Marks

SECTION C (continued)

ANIMAL BEHAVIOUR

1. Many animals use colours and patterns to avoid predation. Some butterflies and moths, for example, have paired circular patterns on their wings.

    It has often been assumed that these wing spots are effective deterrents because they resemble the eyes of the predator's own enemies—the *eye mimicry hypothesis*. An alternative explanation suggests that the off-putting effect arises from visual contrast—the *conspicuous signal hypothesis*.

    In a recent experiment to test these hypotheses, targets made from triangular pieces of card were printed with the following patterns.

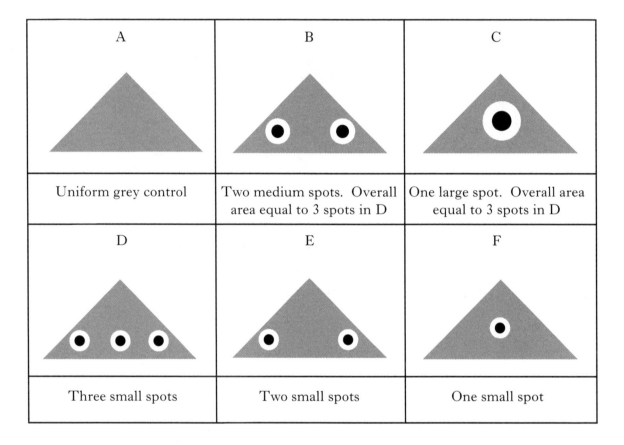

A	B	C
Uniform grey control	Two medium spots. Overall area equal to 3 spots in D	One large spot. Overall area equal to 3 spots in D
D	E	F
Three small spots	Two small spots	One small spot

Each target had a mealworm attached to it. The targets were pinned to trees in woodland containing a variety of predatory bird species. They were checked for predation after 3, 24 and 48 hours. The survival of the mealworms indicated how successful the target patterns had been.

(a) Suggest a variable that should be controlled in the placement of the targets.   1

(b) With which target would there be least predation if the eye mimicry hypothesis is correct?   1

Marks

ANIMAL BEHAVIOUR (continued)

1. **(continued)**

   (c) The table below shows mealworm survival for each target following bird predation.

Target	Survival (%)		
	At 3 hours	At 24 hours	At 48 hours
A	65	8	2
B	95	45	21
C	97	48	20
D	93	35	12
E	88	30	8
F	90	25	6

   Explain how the results support the conspicuous signal hypothesis. 2

   (d) Name **one** other defence strategy where colour or pattern are used to avoid predation. 1

   (5)

2. Describe the benefits obtained by primates living in hierarchical groups. (4)

   [Turn over

Marks

ANIMAL BEHAVIOUR (continued)

3. Three-spined sticklebacks (*Gasterosteus aculeatus*) are fish widely distributed in rivers, lakes, ponds and estuaries throughout the British Isles. During the breeding season, from March to July, the male's throat and belly become a brilliant orange-red and his eyes electric-blue.

(a) In observations of breeding sticklebacks, a checklist of different actions may be used to help in the analysis of their behaviour.

What name is given to such a checklist?   1

(b) The breeding coloration of red throat and blue eye is found in male sticklebacks but not in females.

(i) What term is used to describe this difference between males and females?   1

(ii) State why the male's coloration is important in courtship.   1

(c) A wild, outbred population of sticklebacks was captured and inbred for two generations. The table below shows the effects of inbreeding on some aspects of reproduction in these sticklebacks.

Population	Fertilisation rate (%)	Hatching rate (%)
Wild/no inbreeding	98	94
One generation inbred	95	90
Two generations inbred	84	78

(i) Describe briefly how an inbred population could be produced.   1

(ii) How can these results be explained?   2

(6)

Marks

ANIMAL BEHAVIOUR (continued)

4. (*a*) The graphs below show survival of fruit flies (*Drosophila melanogaster*) from a normal-learning strain and a high-learning strain produced by artificial selection.

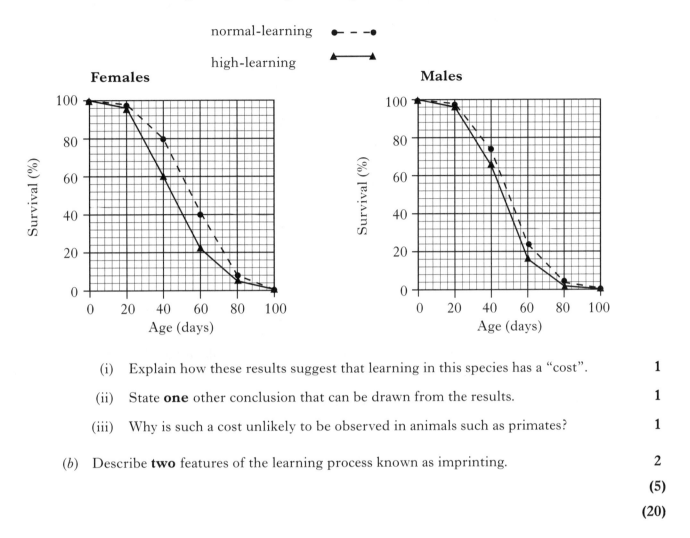

   (i)   Explain how these results suggest that learning in this species has a "cost".   **1**

   (ii) State **one** other conclusion that can be drawn from the results.   **1**

   (iii) Why is such a cost unlikely to be observed in animals such as primates?   **1**

(*b*) Describe **two** features of the learning process known as imprinting.   **2**

   (5)

   (20)

[End of *Animal Behaviour* questions. *Physiology, Health and Exercise* questions start on Page 26]

[Turn over

Marks

SECTION C (continued)

PHYSIOLOGY, HEALTH AND EXERCISE

1. The Figures below show cardiac data for men between 18 and 34 years old who undertake different periods of sports training per week.

Figure 1

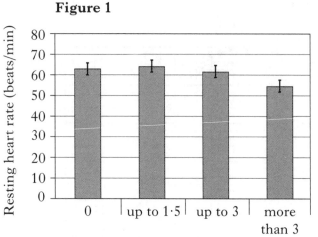

Figure 2

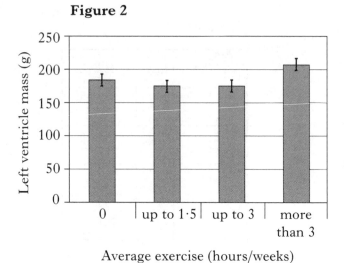

(a) Use information from both Figures to show that the "athletic heart" is an effect of prolonged training. **2**

(b) What other information would be required to determine cardiac output? **1**

(c) What is meant by the term *cardiac hypertrophy*? **1**

(d) Give **one** example of a cardiovascular health benefit that could arise from exercising for less than three hours per week. **1**

(5)

2. Describe the role of insulin in maintaining glucose balance and explain how non-insulin dependent diabetes mellitus (NIDDM) arises. **(5)**

Marks

PHYSIOLOGY, HEALTH AND EXERCISE (continued)

3. The daily intake of calcium for different age groups of females was compared with the recommended intake level. The results shown in the Figure below indicate that actual intake exceeds the recommended level only up to the age of eight.

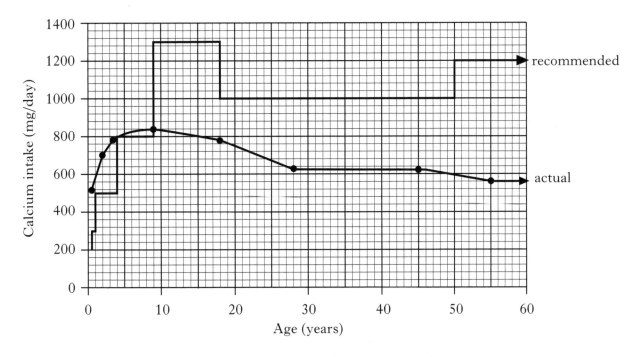

Explain why a high level of calcium intake is recommended for:

(*a*) the 9 to 18 age range;   **1**

(*b*) the over 50s.   **2**

                                                                                           (3)

[Turn over for Question 4 on *Page twenty-eight*

Marks

PHYSIOLOGY, HEALTH AND EXERCISE (continued)

4. (*a*) Direct calorimetry is an accurate but expensive way to measure human energy output.

   Indirect calorimetry uses measurements of oxygen consumption to estimate energy expenditure.

   (i) Describe how energy expenditure is measured by direct calorimetry.   2

   (ii) What **two** aspects of breathing must be measured during indirect calorimetry?   2

   (*b*) It is assumed that a person on a "mixed food" or average diet will release 20·20 kJ of energy for one litre of oxygen used. This energy value is described as the *energy equivalent of oxygen*.

   (i) A typical individual uses 41·5 litres of oxygen walking slowly for an hour.

   How much energy does this activity expend?   1

   (ii) The "average diet" has been worked out so that the energy equivalent of oxygen value it gives is reasonably accurate whatever a person's diet contains.

   The table below lists values for the energy equivalent of oxygen when the diet is made up of a single food.

Food	*Energy equivalent of oxygen* (kJ)
Starch	21·18
Glucose	20·97
Fat	19·67
Protein	19·25

   Use the data to show that the value for an average diet is accurate to within 5% of any diet based on a single food.   2

   (7)

   (20)

[END OF QUESTION PAPER]

[BLANK PAGE]

X007/701

| NATIONAL QUALIFICATIONS 2011 | WEDNESDAY, 1 JUNE 1.00 PM – 3.30 PM | BIOLOGY ADVANCED HIGHER |

SECTION A—Questions 1–25 (25 marks)

Instructions for completion of Section A are given on *Page two*.

SECTIONS B AND C

The answer to each question should be written in ink in the answer book provided. Any additional paper (if used) should be placed inside the front cover of the answer book.

Rough work should be scored through.

Section B (55 marks)

All questions should be attempted. Candidates should note that Question 8 contains a choice.

Question 1 is on Pages 10, 11 and 12. Question 2 is on Page 13. Pages 12 and 13 are fold-out pages.

Section C (20 marks)

Candidates should attempt the questions in **one** unit, **either** Biotechnology **or** Animal Behaviour **or** Physiology, Health and Exercise.

Read carefully

1   Check that the answer sheet provided is for **Biology Advanced Higher (Section A)**.

2   For this section of the examination you must use an **HB pencil** and, where necessary, an eraser.

3   Check that the answer sheet you have been given has **your name**, **date of birth**, **SCN** (Scottish Candidate Number) and **Centre Name** printed on it.

    Do not change any of these details.

4   If any of this information is wrong, tell the Invigilator immediately.

5   If this information is correct, **print** your name and seat number in the boxes provided.

6   The answer to each question is **either** A, B, C or D. Decide what your answer is, then, using your pencil, put a horizontal line in the space provided (see sample question below).

7   There is **only one correct** answer to each question.

8   Any rough working should be done on the question paper or the rough working sheet, **not** on your answer sheet.

9   At the end of the examination, put the **answer sheet for Section A inside the front cover of the answer book**.

Sample Question

Which of the following molecules contains six carbon atoms?

A   Glucose

B   Pyruvic acid

C   Ribulose bisphosphate

D   Acetyl coenzyme A

The correct answer is **A**—Glucose. The answer **A** has been clearly marked in **pencil** with a horizontal line (see below).

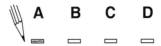

Changing an answer

If you decide to change your answer, carefully erase your first answer and using your pencil, fill in the answer you want. The answer below has been changed to **D**.

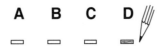

SECTION A

All questions in this section should be attempted.

Answers should be given on the separate answer sheet provided.

1. Cellobiose is a disaccharide with glucose monomers joined by a β 1, 4 bond.

   Which of the following represents cellobiose?

   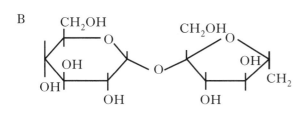

2. The following stages occur during the culture of mammalian cells.

   W   Cells flatten

   X   Cells divide

   Y   Cells become confluent

   Z   Cells adhere to surface

   Which line below shows the correct sequence of stages?

   A   X→W→Y→Z

   B   Z→W→X→Y

   C   Z→Y→W→X

   D   X→Y→Z→W

3. A piece of plant tissue prepared for growth under tissue culture conditions is known as

   A   a callus

   B   an explant

   C   a protoplast

   D   a hybrid.

4. The graphs below show the effect of plant growth substances on the development of roots and shoots in plant tissue culture.

   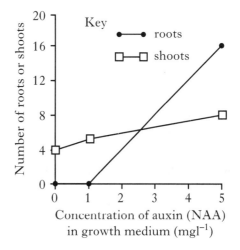

   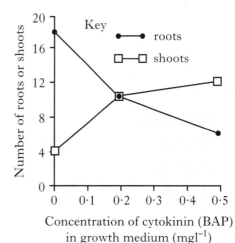

   Which of the following treatments produces a root : shoot ratio of 2 : 1?

   A $0{\cdot}2\,\mathrm{mgl}^{-1}$ BAP

   B $0{\cdot}5\,\mathrm{mgl}^{-1}$ BAP

   C $2\,\mathrm{mgl}^{-1}$ NAA

   D $5\,\mathrm{mgl}^{-1}$ NAA

5. In the formation of fats, which type of linkage is formed by the dehydration (condensation) reaction between glycerol and a fatty acid?

   A   Phosphodiester
   B   Glycosidic
   C   Peptide
   D   Ester

6. Which of the following describes the structure of guanine?

   A   A purine base with a single-ring structure
   B   A purine base with a double-ring structure
   C   A pyrimidine base with a single-ring structure
   D   A pyrimidine base with a double-ring structure

7. An average diploid human cell contains 6×10^9 base pairs of genetic code. Only 1·5% of this may be coding for protein.

   How many base pairs code for protein in a human gamete?

   A $4·5 \times 10^7$
   B $9·0 \times 10^7$
   C $4·5 \times 10^8$
   D $9·0 \times 10^8$

8. Which of the following is responsible for cell-cell recognition?

   A   Glycoprotein
   B   Phospholipid
   C   Hormones
   D   Peptidoglycan

9. The sodium–potassium pump spans the plasma membrane. Various processes involved in the active transport of sodium and potassium ions take place either inside the cell (intracellular) or outside the cell (extracellular).

   Which line in the table correctly applies to the transport of potassium ions?

	Binding location of potassium ions	Conformation of transport protein
A	intracellular	not phosphorylated
B	extracellular	phosphorylated
C	intracellular	phosphorylated
D	extracellular	not phosphorylated

10. Covalent modification can be used to regulate enzyme activity.

    Which of the following is an example of covalent modification?

    A   Allosteric modulation
    B   End product inhibition
    C   Binding of an inhibitor to the active site
    D   Addition of a phosphate group by a kinase enzyme

11. The table below shows the results of an investigation into the effects of varying substrate concentration on the activity of the enzyme phosphatase in the presence of inhibitors. The greater the absorbance the more active the enzyme.

Substrate concentration (%)	Absorbance		
	Inhibitor X	Inhibitor Y	Inhibitor Z
0·1	0·03	0·12	0·06
0·25	0·06	0·17	0·06
0·5	0·14	0·21	0·06
1·0	0·30	0·36	0·06

What valid conclusion can be drawn from the results?

A An increase in substrate concentration reduces the effect of all three inhibitors.

B All three inhibitors are competitive inhibitors.

C Inhibitor Z has least effect on enzyme activity.

D Inhibitor Y has least effect on enzyme activity.

12. A length of DNA is cut into fragments by the restriction enzymes BamHI and EcoRI.

BamHI cut site ▼   EcoRI cut site △

DNA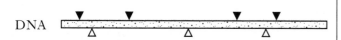

Which of the following gives the correct number of DNA fragments obtained?

	DNA cut by BamHI only	DNA cut by EcoRI only	DNA cut by both BamHI and EcoRI
A	5	4	8
B	4	5	8
C	5	4	9
D	4	5	9

13. The following diagram represents a food web.

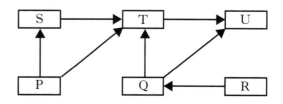

Which line in the table below correctly describes the organisms?

	Producer	Herbivore	Omnivore	Carnivore
A	P	Q	U	S
B	Q	R	S	T
C	R	S	T	U
D	P	U	T	Q

14. A river ecosystem receives about $6\,000\,000\,\mathrm{kJm^{-2}year^{-1}}$ of solar energy. Of this energy 98% is **not** used in photosynthesis.

Which of the following shows the amount of energy $(\mathrm{kJm^{-2}year^{-1}})$ captured by the producers in this ecosystem?

A   120 000

B   588 000

C   1 200 000

D   5 880 000

15. The percentage of energy transferred from one trophic level to the next describes

A   ecological efficiency

B   growth

C   consumption

D   productivity.

[Turn over

16. The Alcon blue butterfly (*Maculinea alcon*) spends most of its life cycle as a caterpillar associated with usually only one species of red ant (*Myrmica* species). Once the caterpillar chews its way out of the flower where the butterfly laid its eggs, it will die unless ants find it. Ants respond to the caterpillar's secretions and adopt it, taking it into their nest. The caterpillar is fed by worker ants and grows quickly, occasionally eating ant larvae.

Which of the following represents the association between the butterfly species and the ant species?

A   Commensal

B   Mutualistic

C   Parasitic

D   Predatory

17. The diagram below represents part of the nitrogen cycle.

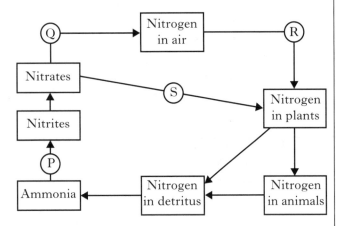

Which of the following stages is likely to involve activity of the enzyme nitrogenase?

A   P

B   Q

C   R

D   S

18. The function of leghaemoglobin is to

A   allow oxygen to react with fixed nitrogen

B   remove oxygen from nitrogen compounds

C   trap nitrogen for use in forming plant proteins

D   trap oxygen to protect bacterial enzymes.

19. The low nitrate content of a marshland soil could result from the activity of

A   *Rhizobium*

B   *Nitrobacter*

C   *Pseudomonas*

D   *Nitrosomonas*.

20. Which of the following is a density-independent effect?

A   An increase in disease decreasing the yield in a crop species

B   An increase in prey numbers increasing the abundance of predators

C   A decrease in grazing increasing the abundance of a plant species

D   A decrease in rainfall increasing the abundance of a plant species

21. The figure below shows two species of butterfly which are bright orange with black markings.

Only the monarch (*Danaus plexippus*) is unpalatable to its predators.

Monarch butterfly

Viceroy butterfly

Which type of mimicry is involved and which species is the mimic?

	Type of mimicry	Mimic
A	Batesian	Monarch
B	Müllerian	Monarch
C	Batesian	Viceroy
D	Müllerian	Viceroy

22. Some insects have a period of dormancy in which a stage of the life cycle is inactive. This type of dormancy is known as

A hibernation

B diapause

C aestivation

D symbiosis.

23. The figure below shows the general relationships between the internal environment and variation in the external environment of four animals.

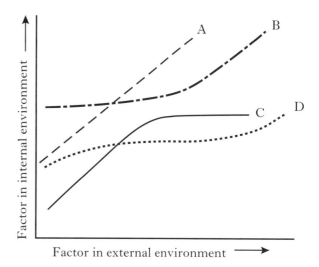

Which animal could occupy the widest range of habitats?

24. Which of the following correctly describes eutrophication?

A Addition of artificial fertiliser to farmland

B Nutrient enrichment in ponds

C Loss of complexity in rivers

D Algal bloom in lochs

[Turn over

25. Cultivation of soya beans is carried out in areas where hedgerows have been cleared to make large open fields.

The crop is regularly treated with herbicides to which soya bean plants are resistant.

Which line in the table below shows changes that would be expected to occur in an ecosystem when a soya bean farm is set up?

	Soil erosion	Species diversity	Density of insect pest species
A	increase	increase	decrease
B	increase	decrease	increase
C	decrease	decrease	decrease
D	decrease	increase	increase

[END OF SECTION A]

Candidates are reminded that the answer sheet MUST be returned INSIDE the front cover of the answer book.

[Turn over for Section B on *Page ten*

SECTION B

All questions in this section should be attempted.
All answers must be written clearly and legibly in ink.

1. *Schistosoma* is a parasitic flatworm found in tropical areas throughout the world. The flatworm can live for many years within a host. In humans, if untreated, it causes the disease schistosomiasis (bilharzia), which can be fatal.

   Schistosoma japonicum is found in East Asia; its life cycle is shown in Figure 1. The parasite's eggs hatch in fresh water, releasing a free-living stage that infects a species of freshwater snail. The parasite multiplies asexually within this secondary host before being released into the water as a second free-living stage. This stage is capable of penetrating the skin of humans and other mammals when they are in fresh water. Inside the liver of the mammal, the flatworms develop into sexually mature adults that disperse eggs via the host's large intestine.

   Figure 1: Life cycle of *Schistosoma japonicum*

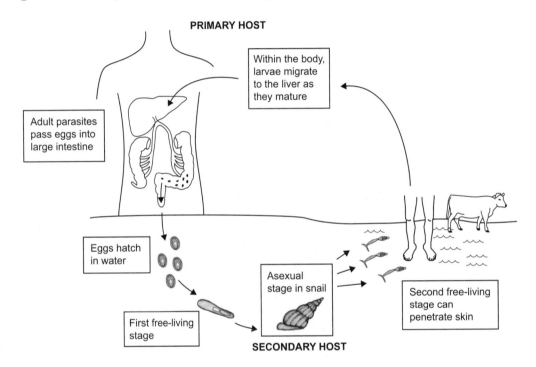

   Successful control of *Schistosoma* is very difficult. Drugs can kill flatworms inside the body but they cannot prevent re-infection. The following factors contribute to high re-infection rates: the parasite has free-living stages; the secondary host can reach high population densities very quickly; untreated, unhygienic humans act as "superspreaders".

   A trial to control *S. japonicum* near a freshwater lake in China compared two pairs of villages given different treatment programmes. Inhabitants of Ximiao and Zhuxi continued to receive the routine annual dose of a drug that kills adult flatworms. Those living in Aiguo and Xinhe were given a programme of intervention that combined the same routine annual drug treatment with the following additional strategies: relevant health education, sanitation, clean bathing water and restrictions on the access of cattle to the lakeside.

   The methods used to evaluate the effectiveness of the intervention programme are shown in the Table and the major findings are shown in Figure 2. The target set for the successful control of *Schistosoma* was to reduce infection in villagers to 1% of the population.

Question 1 (continued)

Table: Methods used to detect *Schistosoma* stages

Stage in life cycle	Detection method for stage
Asexual stage in snail	Dissection of snail samples
Second free-living stage	Dissection of mice exposed in laboratory to samples of lake water
Adult	Non-invasive assessment of human infestation

Figure 2: Infection rates in control and intervention villages

A Control Villages

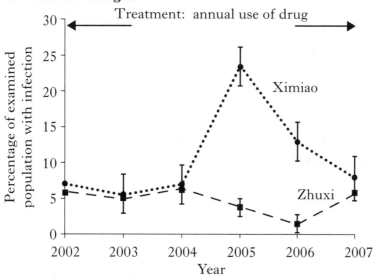

B Intervention Villages

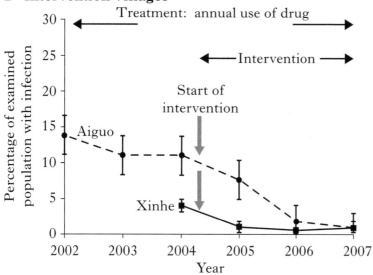

[Question 1 continues on *Page twelve*

Marks

Question 1 (continued)

(*a*)   (i)   Explain what is meant by the term parasitism.   **2**

(ii)   *Schistosoma* has free-living stages but can only feed or reproduce when in contact with a host. What term is used to describe this form of parasitism?   **1**

(*b*)   One of the keys to the successful control of *Schistosoma* is to reduce the number of "superspreaders".

(i)   Suggest a "non-invasive assessment" for identifying superspreaders.   **1**

(ii)   Give **two** aspects of the intervention designed to tackle superspreaders.   **2**

(*c*)   (i)   Use the data from the trial to show that the intervention is needed to achieve the 1% target for infection.   **2**

(ii)   Comment on the reliability of the results.   **1**

(*d*)   Attempts have been made to control *Schistosoma* through the use of molluscicides to kill the secondary host. Using the information, suggest why this method of control is unlikely to be successful with reference to

(i)   the secondary host;   **1**

(ii)   the parasite.   **1**

(*e*)   Explain how the broad host specificity of *S. japonicum* has influenced both the design of the intervention programme and the methods for measuring its effectiveness.   **3**

(14)

[Question 2 is on fold-out *Page thirteen*

Marks

2. The graph below shows variation in global atmospheric carbon dioxide concentration during a fifty year period. Seasonal variation occurs because there is much more green plant biomass in the northern hemisphere than in the southern hemisphere. The underlying trend, however, reveals an increase in concentration.

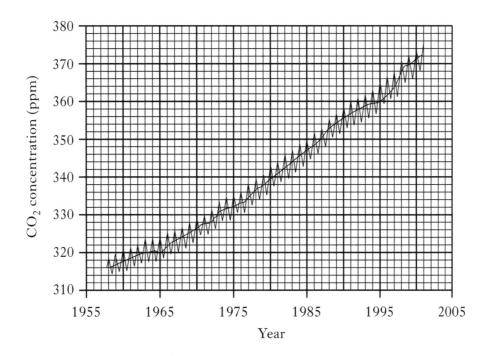

(a) Which **two** cellular processes are responsible for the seasonal variation in carbon dioxide concentration of the atmosphere?   1

(b) Using the trend line, calculate the percentage increase in the carbon dioxide concentration between 1965 and 1995.   1

(c) The increasing carbon dioxide concentration contributes to the enhanced greenhouse effect.

   (i) Explain what is meant by the term *enhanced* in relation to the greenhouse effect.   1

   (ii) Name a gas, other than carbon dioxide, that contributes to the enhanced greenhouse effect.   1

                                                                                                    (4)

Marks

3. Soils that have developed from serpentine rocks have a naturally low abundance of minerals such as calcium, nitrogen, phosphorus and potassium, and a high abundance of potentially toxic metals such as nickel. The succession of serpentine plant communities shows little facilitation and is limited by the regular input of toxic minerals from the erosion of the rock. The climax vegetation that develops tends to be sparse and species present have unusual adaptations to cope with the high concentrations of metal in the soil.

   (*a*) (i) What is meant by facilitation in succession?   **1**

       (ii) What term describes a succession influenced by external factors such as erosion?   **1**

   (*b*) The flowering herb *Alyssum bertolonii* is favoured in serpentine soils because it can isolate absorbed nickel into specialised leaf hair cells. As a result, its dry mass can be as high as 3% nickel. Most other species are susceptible to nickel poisoning at much lower concentrations.

       (i) What term describes the increasing levels of nickel found in *A. bertolonii*?   **1**

       (ii) Explain why a serpentine climax community is unlikely to have many trophic levels.   **1**

       (iii) Suggest why *A. bertolonii* could be used as an indicator species.   **1**

                             (5)

4. Discuss the concept of niche with reference to the competitive exclusion principle.   **(4)**

Marks

5. Myoglobin and haemoglobin are oxygen-carrying proteins. Myoglobin has one polypeptide chain and is found in muscle. Haemoglobin has four polypeptide chains and is found in red blood cells. The tertiary structures of the myoglobin and the haemoglobin chains are very similar. Each chain has one binding site for oxygen.

The proportion of binding sites occupied by oxygen is known as *saturation*.

$$\text{Saturation} = \frac{\text{number of oxygen binding sites occupied}}{\text{total number of oxygen binding sites}}$$

The graph shows the binding of oxygen to haemoglobin and myoglobin as the available oxygen is increased.

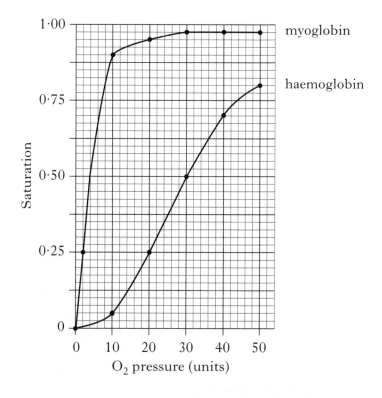

(a) (i) Use data to compare the saturation of myoglobin and haemoglobin between 0 and 30 units. **1**

   (ii) Explain how the information shows that quaternary structure affects the binding of oxygen to haemoglobin. **2**

(b) Use the formula to calculate the change in the number of oxygen molecules bound to haemoglobin as the oxygen pressure is reduced from 30 to 20 units. **1**

(c) Haem groups are an example of non-polypeptide components present in proteins. What term describes such components? **1**

(5)

[Turn over

Marks

6. Gamma-aminobutyric acid (GABA) is a neurotransmitter that functions as a signalling molecule in the central nervous system. GABA binds to a receptor protein located in the plasma membrane of target cells as shown in Figure 1. Binding of a GABA molecule opens a channel that allows chloride ions (Cl^-) to enter the cell.

Figure 1   **Figure 2**

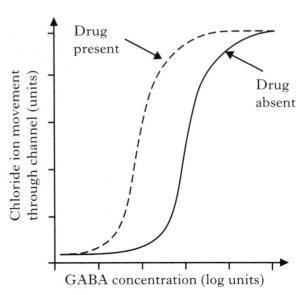

Benzodiazepines are sedative drugs that bind to the receptor protein and increase its affinity for GABA. These drugs act as allosteric modulators by binding at a site that is distinct from the GABA-binding site. Figure 2 above shows the movement of chloride ions through the channel as GABA is increased with and without the drug being present.

(a) (i)   State why neurotransmitters such as GABA cannot cross the membrane.   1

     (ii) What term describes the action of membrane receptors in which signal binding triggers an event in the cytoplasm that alters the behaviour of the cell?   1

(b) (i)   How does the information in Figure 2 show that the affinity of the receptor for GABA has been increased by the drug?   1

     (ii) How might the binding of benzodiazepine to the modulatory site increase the affinity of the receptor for GABA?   1

                                                                                                    (4)

Marks

7. Cystic fibrosis is caused by mutation within the gene encoding the CFTR protein. The most common mutation in this gene is a three base-pair deletion that results in the loss of one amino acid from the CFTR protein. This deletion, ΔF508, accounts for about 70% of mutations in cystic fibrosis.

    A screening test for cystic fibrosis uses the polymerase chain reaction (PCR) to amplify part of the *CFTR* gene containing the mutation.

    (*a*) Describe the features of primers used in PCR.   2

    (*b*) Give **one** technique that could be used in the detection of the mutation following PCR.   1

    (*c*) What information should be given to someone during counselling, following a negative screening result for ΔF508?   1

    (4)

8. Answer **either** A **or** B.

    A. Compare prokaryotic and eukaryotic cells under the following headings:

        (i)   organisation of genetic material;   5

       (ii)   ultrastructure and other features.   10

    OR   **(15)**

    B. Write notes on the cell cycle and its control under the following headings:

        (i)   interphase;   5

       (ii)   mitosis;   5

      (iii)   mutations.   5

    (15)

[END OF SECTION B]

[Turn over for Section C

SECTION C

Candidates should attempt questions on <u>one</u> unit, <u>either</u> Biotechnology <u>or</u> Animal Behaviour <u>or</u> Physiology, Health and Exercise.

The questions on Biotechnology can be found on pages 18–21.

The questions on Animal Behaviour can be found on pages 22–25.

The questions on Physiology, Health and Exercise can be found on pages 26–28.

All answers must be written clearly and legibly in ink.

Labelled diagrams may be used where appropriate.

BIOTECHNOLOGY   *Marks*

1. The figure shows stages involved in the commercial production of an antibiotic.

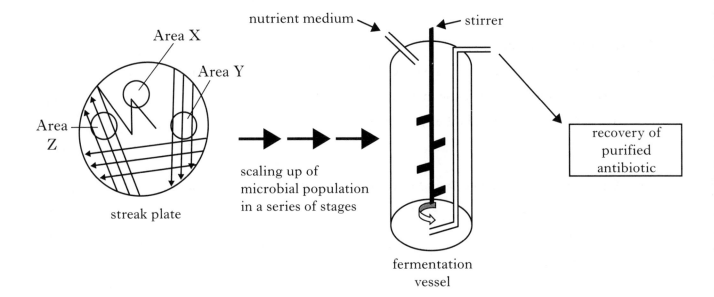

(a) Name a commercially produced antibiotic and the type of micro-organism used in its production.   1

(b) The scaling up process above is started using cells from a single isolated colony.

    (i) Explain why it is necessary to use a single colony isolate.   1

    (ii) Which one of the labelled areas on the streak plate would be most likely to have such a colony?   1

(c) Give **one** reason why the culture in the fermentation vessel is stirred.   1

(d) Give **one** process involved in the recovery of the antibiotic.   1

Marks

BIOTECHNOLOGY (continued)

1. (continued)

(e) The figure shows changes in the culture medium during the production of an antibiotic in a fermentation vessel.

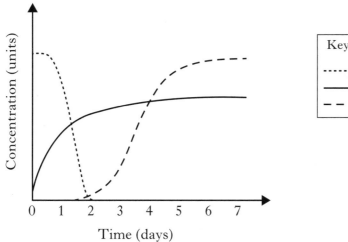

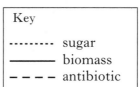

Key

··········· sugar
—————— biomass
– – – – antibiotic

From the figure, give **two** pieces of evidence to indicate that this antibiotic is a secondary metabolite.

2

(7)

[Turn over

Marks

BIOTECHNOLOGY (continued)

2.   The diagram shows stages in the production of monoclonal antibodies.

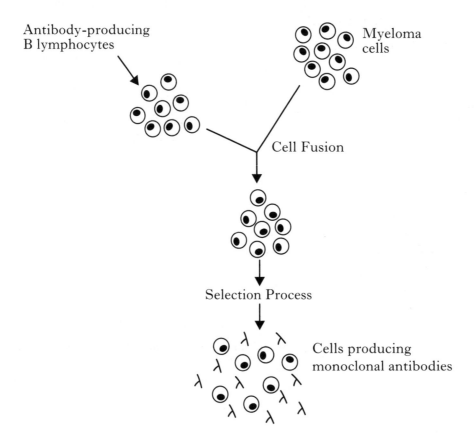

(a)   Describe how the B lymphocytes shown above would have been produced.   2

(b)   Name the chemical used to bring about the fusion of B lymphocytes and myeloma cells.   1

(c)   Give **one** use of monoclonal antibodies in the **treatment** of disease.   1

(4)

3.   The chemical composition of plant cell walls causes problems in the commercial production of fruit juices. Identify the problems and outline how enzymes are used to overcome them.   **(5)**

BIOTECHNOLOGY **(continued)**

Marks

4.   Several studies have demonstrated the antimicrobial activity of oils extracted from plants. One such study investigated the bactericidal activity of an oil from cinnamon bark on the bacterium, methicillin-resistant *Staphylococcus aureus* (MRSA). The oil was added to broth containing an inoculum of MRSA.

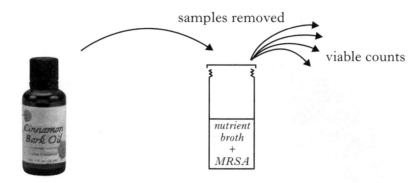

(*a*)   Viable counts were made at intervals over a one hour period.

What is meant by a viable count?   **1**

(*b*)   The graph below shows data for different concentrations of the cinnamon bark oil and a control with no oil. Log units are powers of ten.

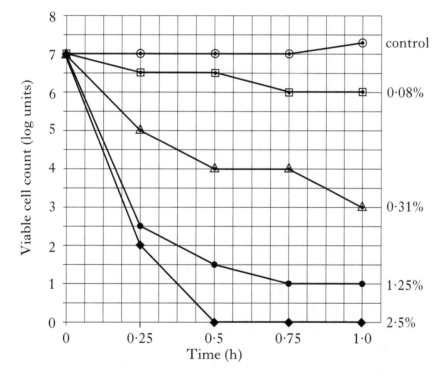

(i)   In this study a bactericidal effect was defined as a reduction, over a one hour period, of 5 log units from the initial viable count.

Identify which of the oil treatments are bactericidal.   **1**

(ii)   What aspect of the procedure was necessary to ensure that a valid comparison was made between the control and the treatments with essential oil?   **1**

(iii)   By how many cells has the starting population been reduced in the 0·08% treatment after the first hour?   **1**

(4)

[End of *Biotechnology* questions. *Animal Behaviour* questions start on Page 22]

(20)

SECTION C (continued)

ANIMAL BEHAVIOUR

1. Northwestern crows (*Corvus caurinus*) can be observed feeding on the beaches of British Columbia in Canada. They search mainly for whelks (*Thais lamellosa*).

Northwestern crow and whelk prey (not to scale)

The crows search only for the largest whelks. After finding a whelk, they take off with it and fly vertically upwards before dropping it onto a rock. This is repeated until the whelk's shell is broken. Steep ascending flight of this kind is energetically expensive. The crows are very persistent and may drop a single whelk up to 20 times before the shell breaks.

(a) The graph below shows the results of an experiment in which **researchers** dropped small, medium and large whelks from different heights. In the graph, "Total height" is obtained by combining the number of drops at each height required to break the whelk. The arrow on the X axis indicates the mean height of drop actually observed when **crows** drop whelks.

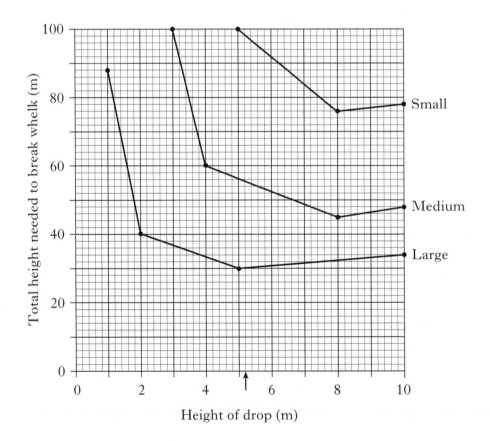

Marks

ANIMAL BEHAVIOUR (continued)

1. (a) (continued)

 (i) On average, how many times would a medium-sized whelk need to be dropped from a height of 4 m in order to break? **1**

 (ii) From the information provided, suggest why crows take only the largest whelks. **1**

 (b) Use the concept of optimal foraging to explain the observation that crows drop whelks from an average height of 5·2 m. **2**

 (c) Apart from energy content and handling time, what other aspect of foraging behaviour is likely to be of significance in determining a predator's choice of prey? **1**

 (5)

2. "All development depends on both nature and nurture." (R. Hinde)

 With reference to life span, compare the roles of nature and nurture in the behavioural development of adult invertebrates and primates. **(5)**

[Turn over

Marks

ANIMAL BEHAVIOUR (continued)

3. The broad-nosed pipefish (*Syngnathus typhle*) can be found along many coasts and estuaries in the British Isles. In this species, "normal" sex roles are reversed: females compete with each other for males and it is the males that are selective. During mating, the female transfers her eggs into a brood pouch in the male where they are then fertilised and nourished. The males provide parental care for the young.

All pipefish are susceptible to infestation by a parasite that induces the formation of visible black spots on the skin.

Figure: Pipefish showing signs of parasite infestation

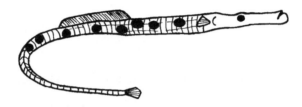

When females are infected, egg production decreases as the parasite load increases. High intensity infections may kill the fish host.

Experiments were carried out to discover male responses to spotted females that were either naturally infected or had been artificially tattooed using ink dissolved in a solvent. In both situations, it was found that males selected females with fewer or no black spots. Behavioural interactions between males, however, were not influenced by the presence or absence of spots on the males.

(*a*) During mating, the fish are more vulnerable to predators.

    (i) Explain how the male's preference for healthy females allows him to maximise his reproductive fitness.   **2**

    (ii) Explain why the genes controlling this behaviour might be described as "selfish".   **1**

(*b*) Give **one** aspect of behaviour that contributes to the high level of parental investment shown by the male pipefish.   **1**

(*c*) The parasite cannot be transmitted directly from one fish to another.

Which aspect of pipefish behaviour is consistent with this observation?   **1**

(*d*) Suggest a control that should be used in the experiment involving tattooed spots.   **1**

(*e*) In the experiments, fish were arranged so that males could see females but females could not see males. Explain why this would strengthen the conclusion that males select against parasitised females using visual stimuli.   **1**

    (7)

Marks

ANIMAL BEHAVIOUR (continued)

4. Cannibalism occurs when animals eat other animals of their own species. Willow leaf beetles, *Plagiodera versicolora*, lay eggs in clutches (groups). Females may have mated once or several times so clutches contain a mixture of half and full siblings. This means that the coefficient of relatedness, **r**, varies between 0·25 and 0·5.

Larvae that hatch first eat unhatched eggs from the same clutch. 24 hours after hatching, cannibals are 30% heavier than non-cannibals.

The scatterplot shows cannibalism rate and clutch relatedness for a number of populations.

Figure: Cannibalism rate versus average clutch relatedness for each of eight populations

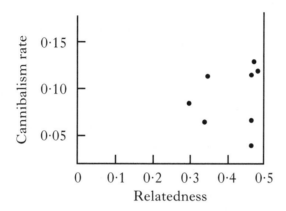

(a) Why do full siblings have a coefficient of relatedness of 0·5?   1

(b) A hypothesis from Hamilton's rule would be that cannibalistic larvae should not eat close relatives.

    (i) What is meant by Hamilton's rule?   1

    (ii) What evidence from the graph contradicts the hypothesis?   1

                                                  (3)

                                                (20)

[End of *Animal Behaviour* questions. *Physiology, Health and Exercise* questions start on Page 26]

[Turn over

SECTION C (continued)   *Marks*

PHYSIOLOGY, HEALTH AND EXERCISE

1. Describe how atherosclerosis can lead to myocardial infarction.   **(4)**

2. (*a*) To reduce the risk of cardiovascular disease, individuals are encouraged to improve the ratio of high-density lipoprotein (HDL) to low-density lipoprotein (LDL).

Give **two** lifestyle factors that can be changed to improve the ratio of HDL:LDL.   2

(*b*) Cholesterol in the blood is associated with HDL, LDL and triglycerides. Treatment with *statin* medication aims to improve the ratio of HDL to LDL by reducing cholesterol production.

The data below show the blood lipid profiles of a patient before and after two years of statin medication, and the normal range of values for a healthy individual.

	Concentration (mmol/l)		
Blood lipid	*Before treatment*	*After treatment*	*Normal range values*
Total cholesterol	8·5	5·5	3·0 – 5·0
HDL	1·9	1·8	1·0 – 2·2
LDL		2·9	2·0 – 3·4
Triglycerides	1·5	1·7	0·3 – 2·5
Total cholesterol/HDL	4·5	3·1	about 3·0

(i) LDL is not measured directly, it is calculated from other values using the formula below.

LDL = (Total – HDL) – (Triglycerides / 2·2)

Calculate the LDL value before treatment.   1

(ii) Select **two** pieces of evidence to show that statin treatment has reduced the risk of cardiovascular disease.   2

(iii) The aim for statin treatment is to increase the proportion of HDL to over 30% of the total. Use the data to show that this has been achieved.   1

    (6)

Marks

PHYSIOLOGY, HEALTH AND EXERCISE (continued)

3. (*a*) Blood glucose concentration increases after a meal.

   Describe the events that bring blood glucose concentration back to normal. **2**

   (*b*) In an *oral glucose tolerance test*, an individual has "impaired glucose tolerance" when results are in the range 7·8 to 11·0 mmol/l. A result in this range is referred to as *pre-diabetic*. If untreated, pre-diabetes leads to Type 2 diabetes (NIDDM).

   The underlying cause of impaired glucose tolerance is insulin resistance.

   (i) Explain why cells become less sensitive to insulin in individuals with insulin resistance. **1**

   (ii) For people with pre-diabetes, explain why there would be a long-term benefit from reducing a high waist to hip ratio. **2**

   (5)

[Turn over for Question 4 on *Page twenty-eight*

Marks

PHYSIOLOGY, HEALTH AND EXERCISE (continued)

4. (*a*) Why does taking part in sporting activities during adolescence reduce the risk of osteoporosis–related fractures in later life?   2

(*b*) The figure below shows the results of a study comparing bone mineral density of groups of women involved in different types of sport. The values show mean percentage difference in bone mineral density between the athletes and a control group who have an inactive lifestyle.

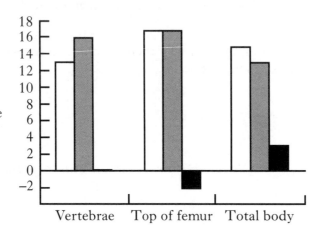

(i) Suggest why the study focuses on bone mineral density of vertebrae and the top of the femur (thigh).   1

(ii) Explain why swimmers are included in the study.   1

(iii) The data suggest that swimming has a negative effect on bone mineral density at the top of the femur. What other information would be required to judge if the results are statistically significant?   1

(5)

(20)

[END OF QUESTION PAPER]

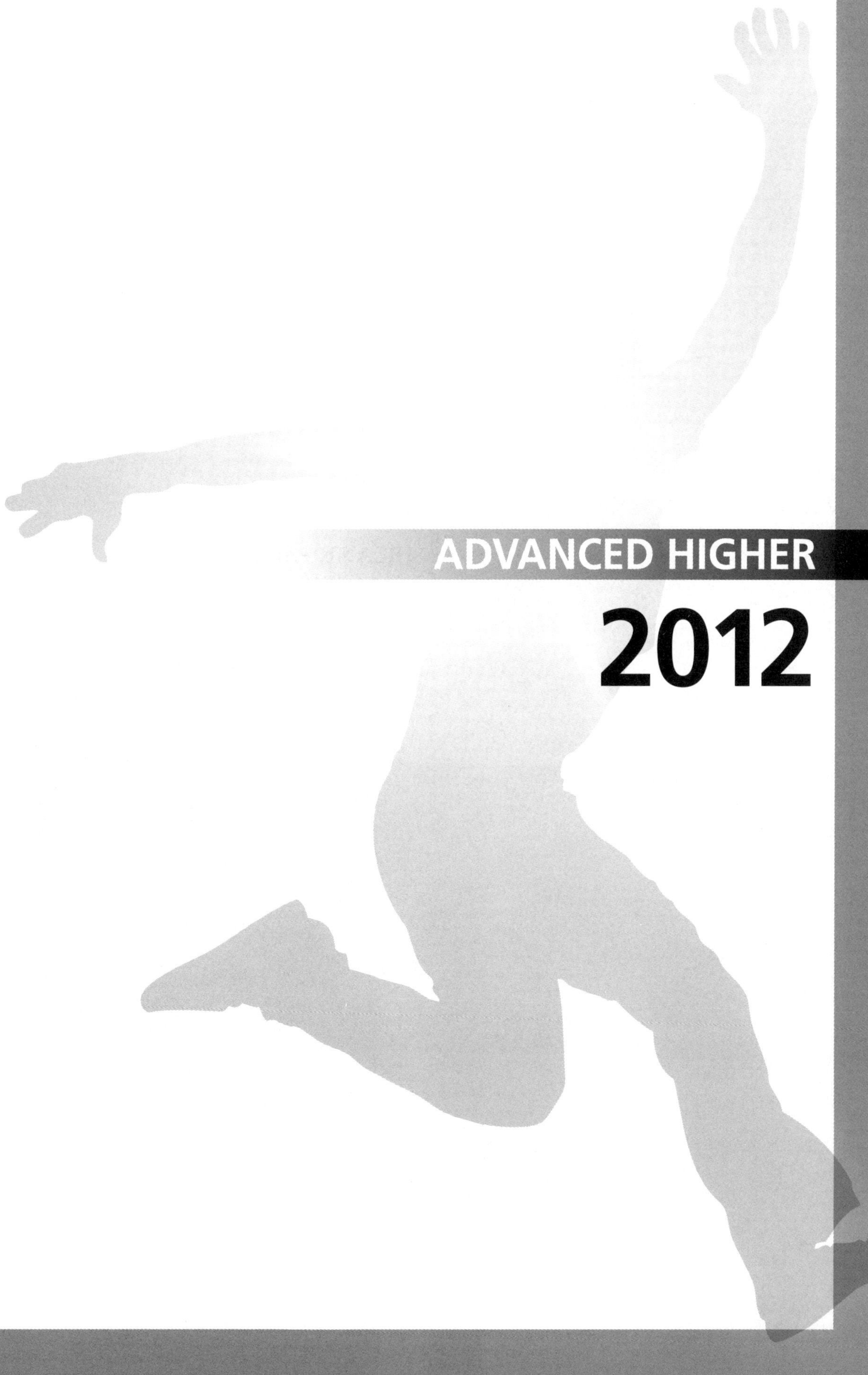

ADVANCED HIGHER

2012

[BLANK PAGE]

X007/13/02

NATIONAL	WEDNESDAY, 23 MAY	BIOLOGY
QUALIFICATIONS	1.00 PM – 3.30 PM	ADVANCED HIGHER
2012		

SECTION A—Questions 1–25 (25 marks)

Instructions for completion of Section A are given on *Page two*.

SECTIONS B AND C

The answer to each question should be written in ink in the answer book provided. Any additional paper (if used) should be placed inside the front cover of the answer book.

Rough work should be scored through.

Section B (55 marks)

All questions should be attempted. Candidates should note that Question 8 contains a choice.

Question 1 is on Pages 10, 11 and 12. Questions 2 and 3 are on Page 13. Pages 12 and 13 are fold-out pages.

Section C (20 marks)

Candidates should attempt the questions in **one** unit, **either** Biotechnology **or** Animal Behaviour **or** Physiology, Health and Exercise.

Read carefully

1   Check that the answer sheet provided is for **Biology Advanced Higher (Section A)**.

2   For this section of the examination you must use an **HB pencil** and, where necessary, an eraser.

3   Check that the answer sheet you have been given has **your name**, **date of birth**, **SCN** (Scottish Candidate Number) and **Centre Name** printed on it.

     Do not change any of these details.

4   If any of this information is wrong, tell the Invigilator immediately.

5   If this information is correct, **print** your name and seat number in the boxes provided.

6   The answer to each question is **either** A, B, C or D. Decide what your answer is, then, using your pencil, put a horizontal line in the space provided (see sample question below).

7   There is **only one correct** answer to each question.

8   Any rough working should be done on the question paper or the rough working sheet, **not** on your answer sheet.

9   At the end of the examination, put the **answer sheet for Section A inside the front cover of the answer book**.

Sample Question

Which of the following molecules contains six carbon atoms?

A   Glucose

B   Pyruvic acid

C   Ribulose bisphosphate

D   Acetyl coenzyme A

The correct answer is **A**—Glucose. The answer **A** has been clearly marked in **pencil** with a horizontal line (see below).

Changing an answer

If you decide to change your answer, carefully erase your first answer and using your pencil, fill in the answer you want. The answer below has been changed to **D**.

SECTION A

All questions in this section should be attempted.

Answers should be given on the separate answer sheet provided.

1. Which line in the table below correctly represents the organelles in a prokaryotic cell?

	Chloroplast	*Mitochondria*	*Ribosomes*
A	Present	Absent	Absent
B	Absent	Absent	Present
C	Absent	Present	Absent
D	Present	Present	Present

2. The following diagram shows a bacterial cell.

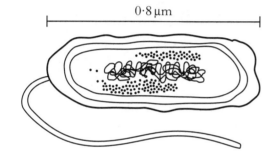

The length of this cell in millimetres (mm) is

A   800

B   80

C   0·008

D   0·0008.

3. In which of the following do both copies of the gene require a mutation for cancer to develop?

A   Oncogenes

B   Proto-oncogenes

C   Proliferation genes

D   Anti-proliferation genes

4. Which of the following is the correct sequence of stages in the production of plants by tissue culture? (PGR = plant growth regulators)

A   callus $\longrightarrow$ explant $\xrightarrow{\text{PGR}}$ plantlet

B   explant $\xrightarrow{\text{PGR}}$ callus $\xrightarrow{\text{PGR}}$ plantlet

C   callus $\xrightarrow{\text{PGR}}$ explant $\xrightarrow{\text{PGR}}$ plantlet

D   explant $\longrightarrow$ callus $\xrightarrow{\text{PGR}}$ plantlet

5. The key below can be used to identify carbohydrates.

1 { Sugars.. go to ...2
    { Polysaccharides .... go to ...4

2 { Monosaccharides................................**A**
    { Disaccharides .... go to ...3

3 { Contains only one type of monomer .... **B**
    { Contains two types of monomer ....sucrose

4 { Storage function .... go to ...5
    { Structural function in plants .... **C**

5 { Storage function in animals......... glycogen
    { Storage function in plants ....**D**

Using the key, which letter would represent amylopectin?

[Turn over

6. Which of the diagrams below represents correctly a molecule of the steroid testosterone?

A

B

C

D

7. Which of the following is correct for a purine base?

	Ring structure	Example of base
A	double	adenine
B	double	thymine
C	single	adenine
D	single	thymine

8. A section of a double stranded DNA molecule contains 80 bases. 24 of these are thymine. The percentage of cytosine bases in the molecule is

A   12

B   16

C   20

D   30.

9. The diagram below shows a small polypeptide integrated into a membrane.

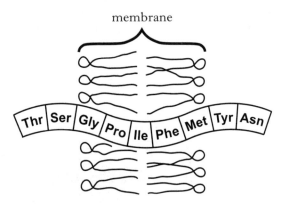

Which line in the table below correctly classifies amino acids in this polypeptide?

	Polar	Non-polar
A	Thr	Pro
B	Ile	Tyr
C	Asn	Ser
D	Phe	Gly

10. The mechanism of action of the sodium-potassium pump involves the following stages:

    P   membrane protein is phosphorylated

    Q   sodium ions bind to membrane protein

    R   sodium ions are released

    S   membrane protein changes conformation

    The correct sequence is

    A   P, Q, R, S

    B   Q, P, S, R

    C   Q, P, R, S

    D   P, Q, S, R

11. The figure below shows how the ATPase activity of the sodium-potassium pump is affected by the concentrations of sodium and potassium ions.

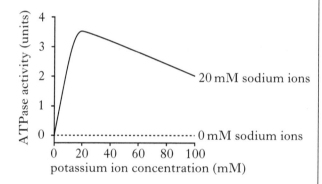

    What valid conclusion can be drawn from this information?

    A   The presence of potassium ions inhibits ATPase activity.

    B   The optimal concentration of sodium ions for ATPase activity is 20 mM.

    C   ATPase activity requires the presence of both sodium and potassium ions.

    D   ATPase activity requires the presence of sodium ions only.

12. The DNA sequences of the normal and mutated versions of a gene are shown below.

    Normal DNA sequence

    GAGAATCCTTGAGCTCTTAAGCTTATT

    Mutated DNA sequence

    GAGAATCCTTGAGGTCTTAAGCTTATT

    The table below gives the recognition sequences of four restriction endonucleases.

Restriction endonuclease	Recognition sequence
BamH1	GGATCC
EcoR1	GAATTC
HindIII	AAGCTT
SacI	GAGCTC

    Which of the restriction endonucleases would produce different numbers of fragments when used to digest normal and mutant DNA?

    A   BamH1

    B   EcoR1

    C   HindIII

    D   SacI

13. During the production of transgenic tomato plants, plasmids can be used to transfer recombinant DNA from

    A   Rhizobium to plant cell protoplasts

    B   Rhizobium to differentiated plant cells

    C   Agrobacterium to plant cell protoplasts

    D   Agrobacterium to differentiated plant cells.

14. When a caterpillar consumes a plant leaf containing 200 kJ of energy, it passes 100 kJ of energy in its faeces, uses 67 kJ of energy for cellular respiration and uses 33 kJ of energy for new growth.

    The amount of energy lost from the woodland ecosystem in this process is:

    A   67 kJ

    B   100 kJ

    C   167 kJ

    D   200 kJ

Questions 15 and 16 refer to the following diagram which shows the annual flow of energy through a terrestrial ecosystem. The units are $kJ\,m^{-2}$.

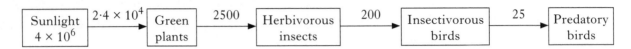

15. The organisms at trophic level 2 are

    A   producers

    B   primary consumers

    C   secondary consumers

    D   tertiary consumers.

16. Gross primary productivity (in $kJ\,m^{-2}$) for this ecosystem is

    A $2\cdot4 \times 10^4$

    B $2\cdot5 \times 10^3$

    C $4\cdot0 \times 10^6$

    D $21\cdot5 \times 10^3$.

17. The graph shows how productivity in a marsh was affected **after a time** by the experimental addition of nitrate and phosphate. Neither was added in the control experiment.

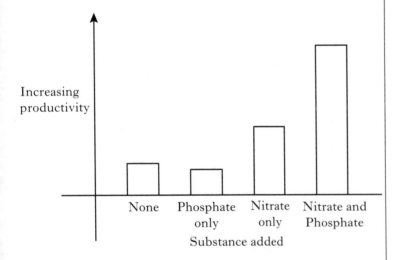

Which statement is supported by the graph?

    A   Productivity in the control is limited by both nitrate and phosphate.

    B   Phosphate can limit productivity if enough nitrate is available.

    C   Phosphate limits productivity in the control experiment.

    D   Productivity in the marsh is never limited by phosphate.

18. The pyramid below represents organisms in a food chain.

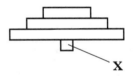

The part labelled **X** could represent

    A   phytoplankton in a pyramid of productivity

    B   oak trees in a pyramid of biomass

    C   phytoplankton in a pyramid of biomass

    D   oak trees in a pyramid of productivity.

19. Which of the following is an example of Batesian mimicry?

    A   Two harmful species of wasps that look like each other.

    B   A butterfly with large eyespots on its wings.

    C   A stick insect which looks like a twig.

    D   A harmless snake which resembles a poisonous species.

20. An experiment was carried out to investigate the density-dependent spread of a fungal disease of plants. Soil samples were taken near an infected plant. Half of the samples were sterilised.

The samples were then used to grow seedlings of the same species of plant at low or high density.

Which line in the table below would result in the highest percentage survival of the seedlings?

	Seedling density	Soil sterilised
A	low	no
B	high	no
C	low	yes
D	high	yes

21. The graphs below show how environmental changes can affect the internal conditions in aquatic organisms W, X, Y and Z.

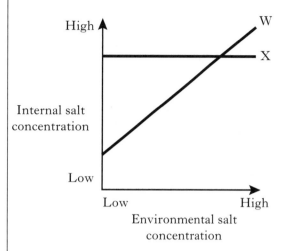

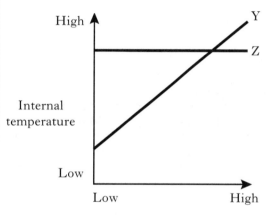

Which line in the table correctly identifies the osmoconformer and the homeotherm?

	Osmoconformer	Homeotherm
A	W	Y
B	W	Z
C	X	Y
D	X	Z

[Turn over

22. The table below shows information about plants in fenced and unfenced grassland plots after 2 years. The plots were fenced to exclude voles which feed mainly on annual grasses.

	Relative biomass (units)		Number of plant species	
	Fenced plots	Unfenced plots	Fenced plots	Unfenced plots
Annual grasses	120	40	6	6
Other plants	40	80	12	24

Which line of the table below best summarises the effects of grazing on the grassland?

	Plant growth (biomass units)	Plant diversity (number of species)
A	increased	increased
B	increased	decreased
C	decreased	decreased
D	decreased	increased

23. Which of the following does **not** result in loss of complexity in ecosystems?

A   Predation

B   Monoculture

C   Eutrophication

D   Toxic pollution

24. Bleaching of coral occurs because

A   pollution prevents them producing coloured pigments

B   toxic chemicals kill zooxanthellae in the corals

C   their symbiotic partners are sensitive to increasing temperature

D   zooxanthellae in the corals are sensitive to UV light.

25. Allogenic succession takes place

A   as a result of climatic change

B   after clearing of agricultural land

C   when more sand is deposited on a beach

D   during decomposition.

[END OF SECTION A]

Candidates are reminded that the answer sheet MUST be returned INSIDE the front cover of the answer book.

[Turn over for Section B on *Page ten*

SECTION B

All questions in this section should be attempted.

All answers must be written clearly and legibly in ink.

1. Recently a new class of RNA, called **micro**RNA, has been discovered. These small RNA molecules have an important role in controlling the translation of mRNA. This type of control is called *RNA interference*.

   A microRNA is formed from a *precursor* RNA molecule that folds into a double-stranded "hairpin" structure. The hairpin is then processed to give a shorter molecule by the enzymes "Drosha" and "Dicer". One strand of this short molecule attaches to RISC proteins; the resulting complex binds to target mRNA molecules and prevents translation. (Figure 1)

Figure 1: Control of gene expression by RNA interference

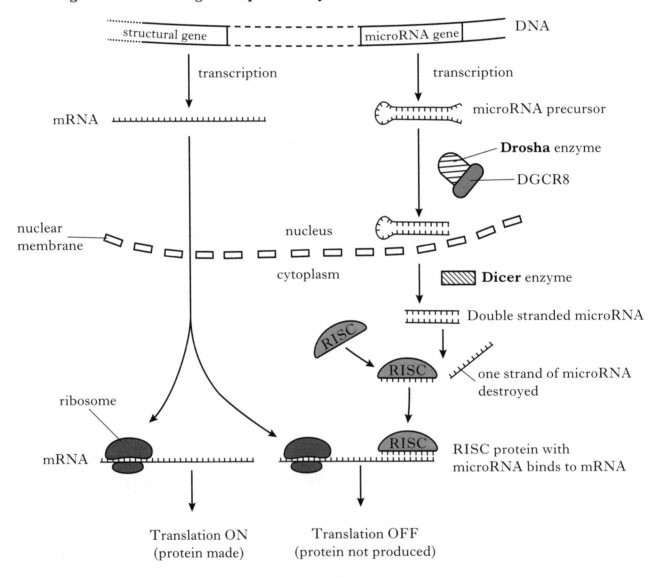

Recent research has investigated the importance of microRNA in controlling the fate of stem cells. Stem cells can either divide rapidly to make more stem cells, a process called **self-renewal**, or differentiate into specialised cell types. To determine the role of microRNAs in these processes, stem cells were modified to "knock out" microRNA production. These microRNA *knockout cells* lack the protein DGCR8, an activator of Drosha. Figures 2A and 2B compare growth rate and cell-cycle progression in knockout and normal cells.

Question 1 (continued)

In further work, the differentiation of knockout and normal cells was studied by inducing the cells to differentiate. Analysis was carried out on the levels of specific marker molecules whose presence is associated with either self-renewal or differentiation. Results are shown in Figures 3A and 3B.

Figure 2A: Effect of knockout on growth rate

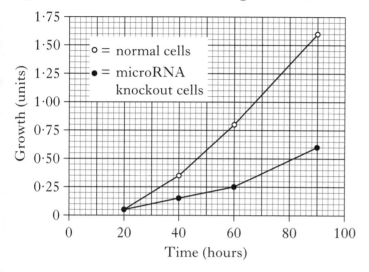

Figure 2B: Effect of knockout on cell cycle

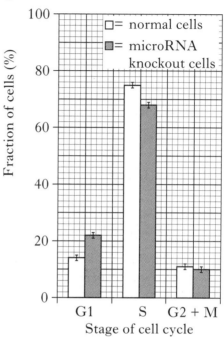

Figure 3A: Level of self-renewal marker

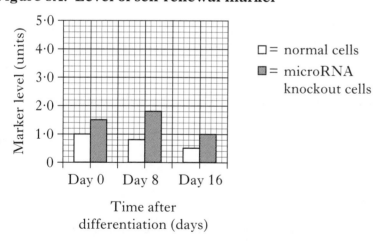

Figure 3B: Level of differentiation marker

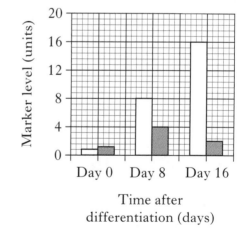

[Question 1 continues on *Page twelve*

Marks

Question 1 (continued)

(*a*) During the formation of microRNAs, single-stranded RNA molecules form hairpin structures as shown in the diagram below.

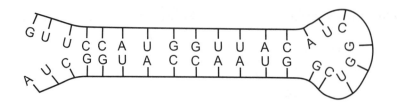

    (i)   Which covalent bonds join nucleotides in RNA molecules?   1

    (ii)   What is the role of hydrogen bonding in maintaining the hairpin shape?   1

(*b*) Describe how the knockout of DGCR8 affects RNA interference.   2

(*c*)   (i)   Refer to Figure 2A. Calculate the percentage reduction in growth at 90 hours caused by the microRNA knockout.   1

    (ii)   The authors concluded that microRNA knockout cells do not progress normally through the cell cycle. How do the results in Figure 2B support this conclusion?   2

(*d*) Refer to Figures 3A and 3B.

    (i)   Comparing normal and knockout cells, give **two** general conclusions about the expression of the differentiation marker.   2

    (ii)   What evidence is there that self-renewal is switched off as differentiation proceeds and that the interaction of these two processes is abnormal in knockout cells?   2

(*e*) MicroRNAs inhibit *translation*. Describe how the *transcription* of β-galactosidase in prokaryotes is switched off.   2

                                      (13)

[Questions 2 and 3 are on fold-out *Page thirteen*

Marks

2. Some fish species can change the colour of their skin by moving pigment granules within skin cells. The granules are attached to microtubules and can either aggregate in the centre of the cell or disperse throughout the cytoplasm.

Movement of granules along microtubules is controlled by hydrophilic signalling molecules that alter the concentration of the intracellular signalling molecule cyclic AMP (cAMP).

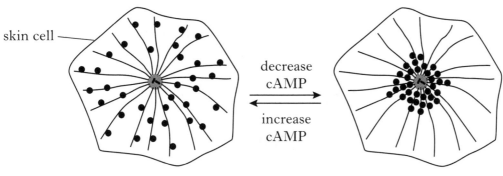

Pigment dispersed   Pigment aggregated

(a) (i) Explain why the control of pigment movement by hormones is an example of signal transduction.   2

(ii) Suggest how the movement of pigments to alter skin colour could function in defence against predation.   1

(b) (i) Microtubules have a role in governing the location of cell components. Give **one** other function of the cytoskeleton.   1

(ii) Name the protein component of microtubules.   1

(iii) From which structure do microtubules radiate?   1

  (6)

3. Describe the control of enzyme activity by competitive and non-competitive inhibitors.   **(5)**

Marks

4. Duchenne muscular dystrophy (DMD) is an inherited condition resulting from a deletion mutation within the dystrophin gene of the X-chromosome.

    A sample of DNA from an individual (P) suspected of having DMD was isolated and digested with a restriction endonuclease. A corresponding control sample (C) was treated in the same way. The resulting fragments were separated using gel electrophoresis. The outcome is shown in the diagram below.

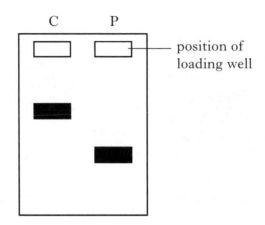

    (a) How are specific gene fragments identified in this procedure?   1

    (b) State whether or not the results confirm a diagnosis of DMD and explain your answer.   2

    (c) An alternative method of genetic screening for DMD involves the amplification of regions of the dystrophin gene. Name the technique used to amplify DNA.   1

                                                                                          (4)

Marks

5. The gannet (*Morus bassanus*) is a fish-eating seabird that breeds on barren, rocky offshore islands in the North Atlantic. A study from 1963 to 1976 investigated the negative impact of DDE on gannet breeding success. DDE, a residue from DDT breakdown, causes thinning of egg shells.

   (a) What term is used to describe the increase in DDE concentration shown in the table below?   1

Gannet age class	DDE concentration in muscle tissue (ppm)
1 year	0·08
2 years	0·50
3–4 years	0·96
Adult	2·17

   (b) The table below shows aspects of breeding success in two different island colonies.

Island colony	Egg hatching success (%)	Chick survival (%)
Ailsa Craig, Scotland	81	92
Bonaventure, Canada	38	78

   (i) At Ailsa Craig, 75% of eggs laid resulted in the survival of a chick.

   Use the data to calculate the corresponding survival rate for eggs laid at Bonaventure.   1

   (ii) DDT was sprayed to protect hillside forests in mainland Canada from severe caterpillar outbreaks.

   Explain how the pollutant came to be present in the gannets at Bonaventure.   2

   (4)

[Turn over

Marks

6. The Antarctic krill (*Euphausia superba*) is the major primary consumer in the marine food web of the Southern Ocean. They feed on algae, the producers. The ecological efficiency of krill is low.

Antarctic krill

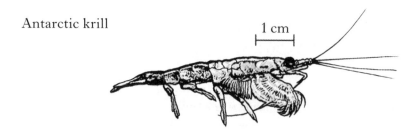

1 cm

(a) Ecological efficiency is the energy in one trophic level as a percentage of energy in the level below. Give **one** reason for a low ecological efficiency.   **1**

(b) Young krill scrape algae from below the ice sheets that form during winter. Once the ice melts, the krill must feed on algae in open water. At this time, they become the main food source of many Antarctic species including penguins, seals and whales.

Long-term studies have monitored krill density in relation to winter ice duration.

The results below suggest that as winter ice duration increases the population density of krill increases.

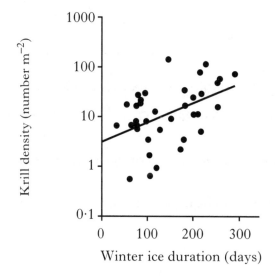

(i) Suggest an explanation for the trend shown in the graph.   **1**

(ii) Krill have the highest total biomass of any species of animal. Their faeces fall to the deep ocean floor where decomposition rates are so low that there is no significant recycling of carbon dioxide back to the atmosphere.

With reference to krill faeces, explain how a rise in sea temperature in Antarctica caused by global warming could lead to a further increase in global warming.   **2**

(4)

Marks

7. The rose-grain aphid (*Metopolophium dirhodum*) is a herbivorous insect that requires two different types of plant to complete its lifecycle. It spends the winter as an egg in diapause on wild roses found in hedgerows. In spring, its numbers build up and it migrates to feed on nearby wheat crops. The rose-grain aphid is an important vector for major plant viruses that reduce grain yields.

   (*a*) What is meant by the term diapause?   1

   (*b*) Suggest why the increased field sizes associated with intensive wheat cultivation may help to reduce crop losses due to rose-grain aphid outbreaks.   2

   (*c*) What is meant by the biological term "vector"?   1

       (4)

8. Answer **either** A **or** B.

   A. Write notes on niche and competition. Use examples as appropriate.   **(15)**

   OR

   B. Give an account of the circulation of nutrients under the following headings:

       (i) decomposition;

       (ii) nutrient cycling.   **(15)**

[END OF SECTION B]

[Turn over for Section C

SECTION C

Candidates should attempt questions on <u>one</u> unit, <u>either</u> Biotechnology <u>or</u> Animal Behaviour <u>or</u> Physiology, Health and Exercise.

The questions on Biotechnology can be found on pages 18–21.

The questions on Animal Behaviour can be found on pages 22–25.

The questions on Physiology, Health and Exercise can be found on pages 26–28.

All answers must be written clearly and legibly in ink.

Labelled diagrams may be used where appropriate.

BIOTECHNOLOGY

1. The diagram shows stages in the production of a monoclonal antibody.

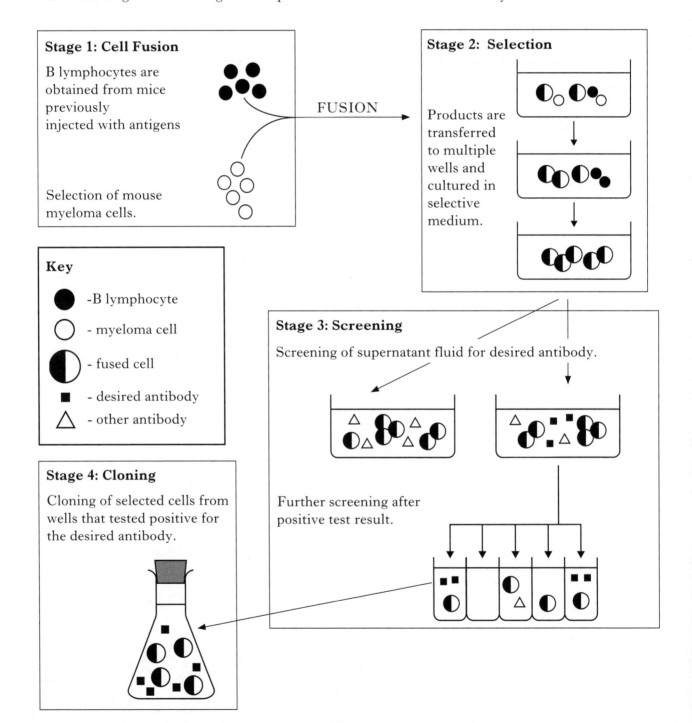

Marks

BIOTECHNOLOGY **(continued)**

1. **(continued)**

    (*a*) Which organ in the mouse is used as a source of B lymphocytes?   **1**

    (*b*) Why are the mouse B lymphocytes fused with myeloma cells?   **1**

    (*c*) Refer to Stage 2 on the diagram.

         (i)   Explain why unfused myeloma cells do not progress to Stage 3.   **1**

         (ii) Explain why B lymphocytes do not progress to Stage 3.   **1**

    (*d*) With reference to Stage 3, explain the need for screening to occur in two steps.   **2**

    (6)

2. Describe how the *growth rate constant* of a bacterial culture can be determined.

    What is its relevance when culturing bacteria for enzyme production?   **(5)**

[Turn over

Mark

BIOTECHNOLOGY (continued)

3. (a) Various enzymes are used in fruit juice production.

   (i) Name an enzyme used to decrease viscosity during extraction.   1

   (ii) Why might arabanase be added to the extracted product?   1

   (b) The diagram illustrates a technique used to purify an enzyme secreted by a culture of microbial cells in a fermenter.

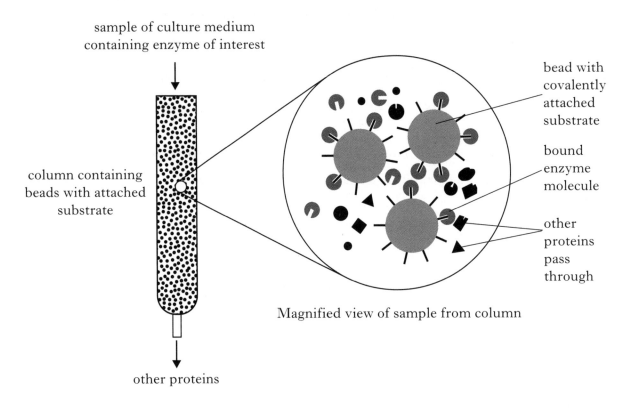

Magnified view of sample from column

   (i) What general term is given to the purification technique shown above?   1

   (ii) What feature of the enzyme molecule allows it to be separated from the other proteins as the sample passes through the column?   1

   (4)

Marks

BIOTECHNOLOGY (continued)

4. To make yoghurt, milk is pasteurised and then inoculated with bacteria that allow fermentation to occur. Two species commonly used together in the inoculum are *Lactobacillus bulgaricus* and *Streptococcus thermophilus*.

    (*a*) What is the purpose of pasteurisation?   1

    (*b*) The fermentation is a two-stage process.

        (i)   State the chemical conversion taking place in the first stage.   1

        (ii) What is the role of the second stage?   1

    (*c*) The graph shows the growth of *L. bulgaricus* and *S. thermophilus* in both pure and mixed culture.

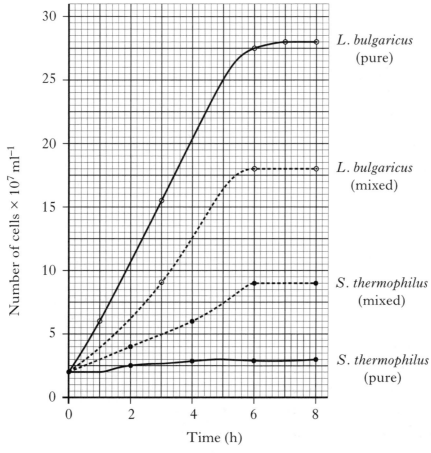

        (i)   Calculate the reduction in growth of *L. bulgaricus* at 8 hours as a result of being in the mixed culture.   1

        (ii) Scientists have suggested that *S. thermophilus* receives a growth promoting substance in this mixed culture. How do the data support that conclusion?   1

                                                                                          (5)

[End of *Biotechnology* questions. *Animal Behaviour* questions start on Page 22]

Mark

SECTION C (continued)

ANIMAL BEHAVIOUR

1. The proportion of time that individual prey animals spend being vigilant may be affected by both the risk of predation and group size. Kudu (Figure 1) are a frequent prey of lions, with most attacks occurring by ambush within two kilometres of water holes. The vigilance behaviour of kudu at water holes has been studied in Hwange National Park, Zimbabwe.

Figure 1: Kudu

Figure 2: Effect of presence of lions and kudu group size on vigilance

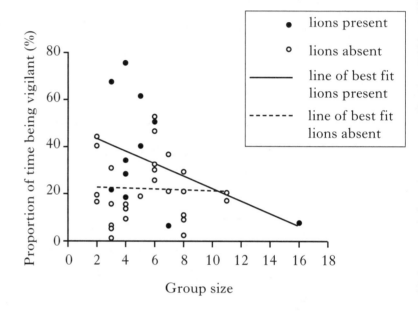

(a) Describe what vigilance behaviour would look like in an animal such as a kudu.   1

(b) Figure 2 shows the proportion of time individual kudu spend being vigilant during drinking, when in different group sizes and when lions are present or absent.

   (i) Suggest **one** strategy that should be used in the observation and recording of kudu vigilance behaviour.   1

   (ii) The researchers hypothesised that individual vigilance would decrease as group size increased, and increase in the presence of lions.

      Use the results to evaluate these hypotheses.   2

   (iii) Comment on the reliability of the data recorded when lions were present.   1

(5)

Marks

ANIMAL BEHAVIOUR (continued)

2. Figure 1 shows the cichlid fish *Cyathopharynx furcifer* found in Lake Tanganyika in Africa. Sexually active males build the sandy substrate into crater-like structures (mating craters). The male fish display around these spawning sites with intense colour. The crater itself has no role in the rearing of the brood produced after fertilisation. Figure 2 shows the results of an investigation into male body size and crater diameter.

Figure 1: Male cichlid in mating crater   **Figure 2: Male body size and crater diameter**

   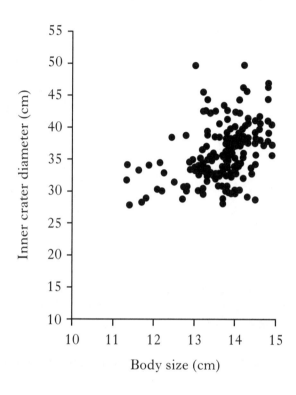

(a) Describe the relationship between body size and crater size shown in Figure 2.   **1**

(b) When crater sizes were either enlarged or reduced by researchers, within a day the males rebuilt the craters to their original sizes, even the enlarged ones.

   (i) Explain why the researchers concluded that mating craters in *C. furcifer* are extended phenotypes.   **1**

   (ii) Give another example of an extended phenotype in a species.   **1**

(c) State **one** feature of a male cichlid that is likely to have evolved as a result of sexual selection.   **1**

   (4)

[Turn over

ANIMAL BEHAVIOUR (continued)

3. Meerkats (*Suricata suricatta*) are social mammals living in the Kalahari Desert. They show co-operative breeding in which a dominant male and dominant female monopolise reproduction. Subordinate animals rarely reproduce but help to rear the offspring of the dominant pair. They are not necessarily closely related to the young animals that they help to rear.

Figure 1: Meerkats

As in the dominance hierarchies of many primate groups, grooming behaviour can be frequently observed in meerkats. Figure 2 shows the frequency of grooming between dominant females and young, subordinate animals.

Figure 2: Grooming of juveniles by dominant females

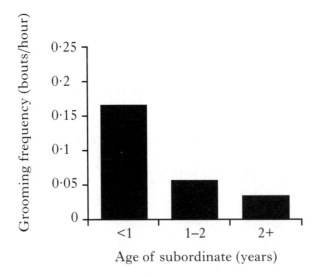

Marks

ANIMAL BEHAVIOUR (continued)

3. **(continued)**

   (*a*) (i) Use Figure 2 to support the hypothesis that one function of meerkat grooming is concerned with parental care.   **1**

   (ii) State an aspect of grooming behaviour, other than frequency, that could be observed and recorded.   **1**

   (*b*) State **two** other functions of grooming in social mammals such as primates.   **2**

   (*c*) Select a statement from the information given about meerkats that may seem to be at odds with the concept of the "selfish" gene. Justify your answer.   **2**

   (6)

4. Discuss the characteristics of sign stimuli and fixed action patterns. How do they interact in the feeding of young birds by their parents?   **(5)**

[End of *Animal Behaviour* questions. *Physiology, Health and Exercise* questions start on Page 26]

[Turn over

Mark.

SECTION C (continued)

PHYSIOLOGY, HEALTH AND EXERCISE

1. In the cardiac cycle, ventricles contract and force blood into the arteries during systole; during diastole the chambers are relaxed and the ventricles fill with blood.

    (*a*) What do the values 120/70 represent in a normal blood pressure reading?   1

    (*b*) Explain how plaque formation in artery walls can lead to raised blood pressure.   2

    (*c*) The most widespread cardiovascular disease in western countries is atherosclerosis in coronary arteries. The most common symptom is angina pectoris, chest pain that develops from oxygen shortage in the myocardial circulation during exertion.

    Oxygen is delivered to heart muscle during diastole as ventricles relax. When heart rate changes, the durations of systole and diastole change, as shown in the table below.

Heart rate (bpm)	Duration of systole (s)	Duration of diastole (s)
65	0·27	0·65
75	0·27	0·53
200	0·16	0·14

    (i) Calculate the % decrease in duration of systole when heart rate increases from 65 to 200 bpm.   1

    (ii) Use information provided to explain why, in individuals with atherosclerosis, exertion can cause angina.   2

    (*d*) Most of the treatments to relieve angina aim to dilate the coronary arteries. However, a relatively new drug, *ivabradine*, has been successful in the treatment of angina by only reducing heart rate.

Marks

PHYSIOLOGY, HEALTH AND EXERCISE (continued)

1. (*d*) (continued)

The figure below shows the effect of exercise tolerance tests on angina patients, exercised until the onset of chest pain. Patients were given tablets containing a dose of *ivabradine* or a placebo (where the tablet contained no drug). The change in response is calculated from each individual's result before and after taking the tablets.

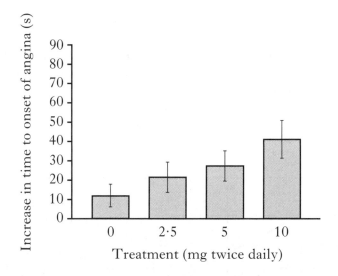

What evidence is there that the medication is effective?   1

(7)

2. Obesity is defined as a body mass index (BMI) greater than 30 (kg/m^2). BMI is not a measure of body composition. Other methods are used to determine percentage body fat, such as densitometry and bioelectrical impedance analysis.

   (*a*) Is an individual with a height of 1·82 m and mass of 90 kg obese?

   Justify your answer.   1

   (*b*) (i) What **two** measurements are required in densitometry?   1

   (ii) How is the density value used to obtain percentage body fat?   1

   (*c*) What is the drawback of using bioelectrical impedance analysis with obese individuals?   1

(4)

[Turn over

Marks

PHYSIOLOGY, HEALTH AND EXERCISE (continued)

3. Medical scientists are increasingly focusing research on *"metabolic syndrome"*, a group of risk factors that apply to both Type 2 diabetes (NIDDM) and cardiovascular diseases. Risk factors include increased insulin concentrations, increased fasting glucose, increased blood triglycerides and decreased HDL. The underlying concern is their common link to *insulin resistance*, a loss of sensitivity to insulin.

    (a) What effect does exercise have on the lipid profile of blood?   1

    (b) Explain how exercise reduces blood glucose in Type 2 diabetes.   2

    (c) In an investigation, volunteers with metabolic syndrome were monitored following different periods of jogging on a treadmill at 60% of their VO_{2max}. They fasted for 12 hours after the exercise period then consumed a high-energy drink and remained at rest. Blood samples were taken following the high-energy drink and at two-hour intervals. Results for insulin concentration are shown below.

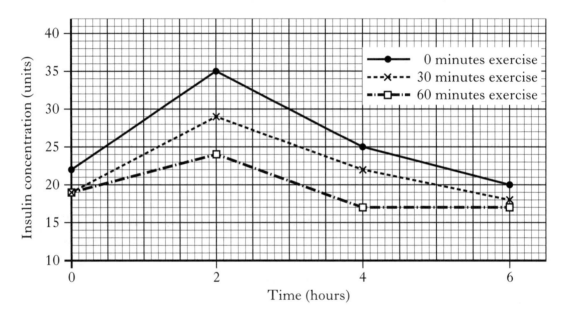

Give **one** conclusion that can be drawn from the experiment about the possible role of exercise in the control of metabolic syndrome. Use data to support your answer.   2

    (5)

4. Discuss the contribution of exercise to a weight-control programme.   **(4)**

[END OF QUESTION PAPER]

[BLANK PAGE]

Acknowledgements

Permission has been sought from all relevant copyright holders and Bright Red Publishing is grateful for the use of the following:

An extract from 'From structure to disease: the evolving tale of aquaporin biology' by Landon S. King, David Kozono and Peter Agre, reprinted by permission from MacMillan Publishers Ltd: *Nature Reviews Molecular Cell Biology*, copyright © 2004 (2008 page 10);

A table adapted from 'Diffusional water permeability of human erythrocytes and their ghosts' by J Brahm, © Brahm 1982. Originally published in *The Journal of General Physiology 79* 791–819 (2008 page 10);

A picture from *Physiological Mini-Reviews* Vol 1, No 10, May 2006 by A Guttierrez. Published by The Argentine Physiological Society (2008 page 10);

A picture from 'Severely Impaired Urinary Concentrating Ability in Transgenic Mice Lacking Aquaporin-1 Water Channels' by Tonghui Ma, Baoxue Yang, Annemarie Gillespie, Elaine J. Carlson, Charles J. Epstein and A. S. Verkman. Published in *Journal of Biological Chemistry* 20 February 1998. Copyright © 1998, by the American Society for Biochemistry and Molecular Biology (2008 page 11).

ADVANCED HIGHER | ANSWER SECTION

BIOLOGY ADVANCED HIGHER
2008

SECTION A

1. C	14. A
2. D	15. D
3. C	16. D
4. D	17. C
5. B	18. A
6. A	19. A
7. D	20. B
8. A	21. C
9. B	22. B
10. A	23. B
11. B	24. A
12. D	25. C
13. A	

SECTION B

1. (a) (i) integral/intrinsic
   (ii) quaternary has sub-units **and** aquaporin has four

   (b) (i) so direction/movement can be traced/ tracked/determined
       or can tell which/how many water molecules came from inside
   (ii) (in isotonic conditions) water molecules will move inwards and at equal volume/rate
   (iii) kinase
   (iv) 1·8 to 20 **or** 18·2 increase (in hypertonic) 18·2/1·8 × 100 = **1011**%

   (c) (i) significant difference between NN(Nn) and nn for loss
       or nn loses 14% more than NN(Nn)
       NN has AQP1 **and** nn doesn't
       or (in the absence of AQP-1) nn unable to reabsorb as much as NN (in descending limb)
   (ii) (when dehydrated) Nn loses same **mass** as NN (21 to 22%) (Fig 3)
       Increase/change in urine concentration is the same/1700 in both NN and Nn
       Reference to error bars overlapping **or** differences not significant

   (d) (i) NDI people will have higher urine output/more dilute urine
       or NDI people will have unchanged urine output following water shortage
   (ii) (normal 70% reabsorption in the descending limb through AQP1 but) no/less reabsorption in collecting duct because AQP2 non-functioning/AQP2 does not reabsorb water/lack of sensitivity to ADH

2. (a) (i) Nucleosome
   (ii) *Any two from:*
       DNA negatively charged + protein positively charged
       or protein neutralises DNA charge
       Compact 'storage' (of DNA) leads to chromosome formation/essential for mitosis

   (b) (i) *Agrobacterium* (*tumefaciens*)/*A. tumefaciens*
   (ii) cellulase and digest cell wall/cellulose

3. 1. four rings **or** correct shape drawn
   And any three from:
   2. variation in side groups in different steroids
   3. hydrophobic/non-polar/lipid-soluble **and** can diffuse/pass/travel through plasma membrane/into cell
   4. bind to proteins/ in cytoplasm/nucleus
   5. (protein complex) regulates genes **or** switches on gene (transcription)
   6. (steroid) hormones (are signalling molecules) eg testosterone

4. (a) Enzyme/active site changes shape **when substrate binds** (to improve the fit)
   (b) it has sites away from the active site where inhibitors/activators/modulators can bind
   (c) AMP inhibits the enzyme **or** AMP exhibits (end-)product inhibition so there is less PRPP/ intermediates **or** the production of AMP decreases **or** AMP 'causes' negative feedback

5. (a) (175/13000) × 100 = 1·3%
   (b) 70 kJ m⁻² day⁻¹
   (c) (Eco efficiency for) P/1 is 12/175 = 6·9%
       or
       (Eco efficiency for) 1/2 is 2/12 = 16·7%
   (d) Heat
   (e) Decomposers external and detritivores internal enzyme digestion

6. (a) Must have oxygen (to carry out metabolism/ respiration)
       or Can only grow in the presence of oxygen.
   (b) (i) *Rhizobium* (in legumes)
   (ii) Nitrogenase
   (iii) Absorbs/binds oxygen to create anaerobic conditions
        or prevents enzyme being inhibited by oxygen
        or prevents oxygen reaching the enzyme
   (c) no effect/no loss/remains the same (since nitrate conversion to N gas/denitrification is anaerobic)

7. (a) Cultivation/growing of single species/crop (over large area).
   (b) If plants are at least **1·25m**/more than **1·25m** apart (no new pustules are formed)
   (c) Fungicide/pesticide
       Selective breeding for resistant varieties
       Genetic engineering/transgenic plants to get resistant varieties

8. (a) (i) Phosphate enrichment:
       Any five from:
       1. Phosphate is a limiting factor in (aquatic) ecosystems
       2. Eutrophication as phosphate/nutrient enrichment
       3. Appropriate source; fertilisers/leachate/ sewage etc.
       4. Algal bloom/algal population explosion mentioned
       5. Growth unsustainable/death of algae occurs **or** less light for plants below
       6. Bacterial decomposition of dead algae /plants **or** toxin production
       7. Oxygen depletion as consequence of bacterial action
       8. Loss of diversity/death of aquatic organisms

   (ii) Exotic species:
      Any four from:
      9. Defined as foreign/introduced/alien species
      10. Example(s)
      11. Reason for success
      12. Description of damaging effect
  (iii) Persistent toxic pollutants:
      Any six from:
      13. Definition of pollution as discharge of harmful substances
      14. Example; eg heavy metals/DDT
      15. Origin of pollutant, eg industry/ agriculture
      16. Non-biodegradable nature/persistence explained
      17. Bioaccumulation = build up in single trophic level
      18. Biological magnification defined as build up in successive trophic levels
      19. Consequences show up at higher trophic levels
      20. Effect described; eg eggshell thinning/reduced disease resistance/death of top predators

8. (b) (i) Dormancy:
      Any seven from:
      1. Period of suspended/reduced activity **or** reduction in metabolism
      2. Means of resisting/tolerating/ surviving/avoiding adverse conditions
      3. Predictive anticipates/before onset of conditions
      4. Consequential is in response to/after onset of conditions
      5. Resting spores/seeds are structures that germinate when suitable conditions return
      6. Diapause - suspended development in insects
      7. Hibernation referring to energy conservation/reduction in body temperature in cold period
      8. Aestivation is response to high T/drought
      9. Named example in context
  (ii) Mimicry:
      Any four from:
      10. Batesian where palatable/harmless species resembles a harmful one
      11. Relative numbers – mimics lower than model
      12. Mullerian where harmful species resemble each other
      13. Example of either type of mimicry; coral snake – false coral; monarch and viceroy butterflies
      14. Reference to aposematic/warning coloration
      15. Survival value = defence against predation
  (iii) Mutualism:
      Any four from:
      16. Mutualism is a form of symbiosis
      17. Mutualism is close/intimate/long-term relationship
      18. Relationship in which both **species** benefit
      19. Name of two species involved
      20. Description of mutual benefits

SECTION C

Biotechnology

1. (a) *Any two from:*
    1. small number of cells to start with/for division
    2. period of adaptation/adjustment
    3. some cells may be depleted of essential growth metabolites/substances/may need to absorb nutrients
    4. cells undergoing repair (due to damage during storage)
    5. essential enzymes are being induced/ change takes place in cells

  (b) (i) overlapping cells are only included in the count once **or** if included overlapping cells lead to overestimate

    (ii) volume over grid = $0·2 \times 0·2 \times 0·1$
        count is 20 cells, so 1 mm^3 contains 20/0·004
        = **5000** cells

  (c) (i) bacteriostatic – inhibitory/cells arrested, not dead **or** regrow when diluted
        bactericidal – lethal/kills
    (ii) Streptomycin/tetracycline/erythromycin etc.

  (d) (i) B lymphocytes
    (ii) Diagnostic testing/ diagnosis of disease
        Elisa
        Treatment of disease (eg rabies, breast cancer)

2. *Any four from:*
    1. polygalacturonase breaks down pectin
    2. which causes fruit to soften **or** modification/treatment slows down ripening/bruising
    3. gene for enzyme is cloned
    4. gene inserted into plant genome in reverse/antisense technology idea
    5. mRNA strand from inserted gene is complementary to mRNA for enzyme **or** sense and antisense mRNAs bind
    6. gene silenced/translation prevented
    7. enzyme not made/reduced enzyme level

3. (a) (i) Trend: as proportion of BK1 increases biomass increases
        + any one correct quantification
    (ii) higher nitrogen content of BK1 or more amino acids available for growth

  (b) Starter for silage **or** for silage production

4. (a) breakdown of yeast cells by their own enzymes

  (b) *Any two from:*
    stage of life cycle
    age of yeast culture
    yeast strain
    pre-treatment of yeast cells with enzymes
    mechanical disruption of yeast cell wall
    pH
    temperature
    enzymes produced by yeast due to genetic makeup

  (c) source of vitamins (as in Marmite)
    flavourings (of soups/gravies/sauces/foods with non animal origin)

Animal Behaviour

1. (a) Female choice
    or advantage following female preference, eg advantage from size through genes **or** burrowing **or** survival
    or
    Male-male rivalry
    or
    Competitive advantage arising from size eg access to more females

  (b) Proximate: to avoid drying out/heat/environmental conditions
    Ultimate: increased breeding success/survival of genes

  (c) Female: greater energy/costs because of bigger gametes/incubation of eggs/carrying young in pouch
    or converse for males

  (d) Genetically determined/stereotyped/instinctive

  (e) Reduced

2. (a) (Natural selection favouring) behaviour that increases relative's survival/fitness (*b*) (Those without B have) no receptors/protein **and** either cannot identify intruder ants **or** don't distinguish other colony members from their own so not aggressive to intruders

or don't maintain loyalty to their queen

(*c*) *Drosophila per* gene/hygienic behaviour in bees

3. *Any four from:*
1. High levels of homozygosity in inbred populations
2. Heterozygosity in outbred populations
3. Inbreeding increases expression of disadvantageous/ lethal recessive genes **or** converse for outbreeding
4. Inbreeding results in lower fitness/inbreeding depression/reduced breeding success **or** converse for outbreeding
5. Male dispersal in mammals/or example
6. Female dispersal in birds/or example

4. (*a*) (i) 1 year versus 3/4/adult,
2 year versus 4/adult
3 year versus adult

(ii) increasing feeding rate as they get older **because** they learn what is edible/compete better/feed more efficiently

(iii) 300 s
or
298·8/299

(*b*) Allow (more) accurate recording (of age/feeding rate)**or** reduces errors
Watch again to check/ensure accuracy
Observer does not affect behaviour

(*c*) Changes in:
diet/or example **or** foraging behaviour/or example habitat preference/or eg **or** nesting behaviour

Physiology, Health and Exercise

1. (*a*) Coronary arteries/coronary blood vessels
(*b*) Build up of plaque/atheroma in vessel wall
Composition of atheroma: fatty material/ cholesterol/fibrous material/smooth muscle cells/calcium loss of elasticity/hardening of vessel walls (increasing BP)/LDL transport of cholesterol for deposition
(*c*) (i) Lumen volume increases (so increased blood flow)
Quantification: doubles lumen volume/65 to 129-133 mm^3
or Atheroma decrease is 15 mm^3
(ii) Angina/Angina pectoris

2. (*a*) Bones become (more) porous/brittle
or reference to spine curvature
(*b*) More common in women with low/reduced oestrogen/in menopausal or post-menopausal women
(*c*) (i) The **number** of women with osteoporosis is increasing (from 7·8 to 10·5 million)
(ii) The **percentage/proportion** of women with low bone mass who go on to develop osteoporosis is decreasing (35·7% to 34·5%.)(ratio 2·8:1 to 2·9:1)
(*d*) Jogging is weight-bearing exercise **or** swimming is not
Jogging/regular weight-bearing exercise maintains/ increases/promotes bone mass/ strength/density **or**
Jogging when younger maximizes bone density before age-related loss

3. (*a*) (i) $(10 \times 29·4)/3 = 98$ (days)
(ii) $96/(1·74 \times 1·74) = 31·7$
(*b*) Increased body mass may be due to a large muscle mass or bone mass **and**
High lean: fat is healthier/muscle accumulation healthy
or high fat: lean unhealthy/fat accumulation is unhealthy

(*c*) Densitometry/skinfold thickness/bioelectric impedance analysis /waist to hip ratio/mid-upper arm circumference.

4. 1. Testing can be maximal or sub-maximal
2. Sub-maximal testing used to monitor/improve aerobic fitness of cardiac patients/less fit
3. Maximal testing measures maximal oxygen uptake/VO$_{2\,max}$ in athletes/fit people
4. Sub-maximal tests are predictive/estimates of VO$_{2\,max}$
5. Predictive/sub-maximal tests assume a linear relationship between heart rate and O$_2$ consumption/exercise intensity
6. Description of exercise stress test(s) used: step test/shuttle test/treadmill **or** Maximal tests to exhaustion

BIOLOGY ADVANCED HIGHER 2009

SECTION A

1. A	14. A
2. B	15. D
3. B	16. C
4. D	17. B
5. B	18. D
6. A	19. A
7. C	20. C
8. D	21. B
9. C	22. A
10. A	23. C
11. D	24. B
12. C	25. D
13. B	

SECTION B

1. (a) An organism that consumes complex/organic molecules
   or consumer
   or feeds on/gets energy from other organisms

   (b) (i) Wolves cause (herbivores decrease) aspen increase
       wolf **increase** data (dates or numbers)
       and corresponding aspen data (damage decrease or
       height increase)
   (ii) 208%
   (iii) *Any one from*:
       • high reproduction rate of elk
       • elk reproduction replaces predation loss
       • prey consumed were not reproductive
       • (young) wolves eat other animals
       • many wolves are young and not hunting yet
       • more vegetation/food for elk so more survive

   (c) *Any two from*:
       Older trees more (abundant) in Zion Canyon than in N
       Creek **or** not much difference
       Younger trees **much** more (abundant) in N Creek than in
       Zion Canyon
       More older trees than young trees in Zion Canyon
       or More younger trees than older trees in North Creek

   (d) *Any one from*:
       • investigators would scare them away/cougars go into
         hiding (because cougars are sensitive to humans)
       • same cougar may be counted twice
       • sightings not reliable/accurate (so counting scats is better)
       • scats more easily spotted than cougars/static
       • numbers of scats proportional to number of cougars

   (e) In Zion Canyon/where there are tourists, cougars are absent
       and there are few saplings/cottonwood
       or
       In North Creek/where there is no tourism the cougars are
       present and there are many saplings/cottonwood
       Tourism started in 1930s (in Zion Canyon) and since then
       there are few trees

   (f) (i) Density dependent
       (ii) *Any two from*:
         Excessive/high intensity grazing leads to less diversity
         because all/most/several/some species are killed
         Moderate grazing increases diversity because dominant
         species are kept in check/others can now grow/rarer
         species survive better with less competition/those with
         basal meristems survive etc
         Low intensity grazing reduces diversity because a
         single species dominates/outcompetes the others

2. (a) (i) Resisting/tolerating/surviving environmental adversity
       (ii) *Any one from*:
         • reduction in population of native species
         • extinction
         • reduction in biodiversity
         • competitive exclusion
         • loss of habitat
         • spread of disease to native plants
       (iii) Herbicide/pesticide/cutting/removal/
         disease/herbivore
       (iv) Need to control for 2 or so years to allow dormant
         seeds to germinate
         or
         Cut down/spray plants before flowering for two years
         or
         Remove/kill all dormant seeds after vegetation removed

   (b) Reduction in soil quality/erosion/compaction
       Reduction humus/nutrients/fertility

3. 1. CO_2 from burning/use of fossil fuels
   2. **enhanced** greenhouse effect
   3. causes global warming
   4. <u>zooxanthellae</u> and coral (symbiosis) is mutualistic
   5. as sea temperature increases algae leave the coral
   6. coral bleaching

4. (a) Harmful for both species/organisms involved (in
       interaction)

   (b) Resources a species is capable of using in **the absence of
       competition**

   (c) *Any two from*:
       No two species with same niche can coexist
       Interspecific competition is intense (because both species
       have similar fundamental niche)
       Competitive exclusion has occurred **or** description of **local**
       extinction of one of the species

   (d) (i) Co-evolution of **host and parasite**
         or
         Specificity of **host and parasite**
       (ii) (Direct) contact/vector/resistant stage/secondary
         host/vertical

5. (a) (i) B, C, D, E
       (ii) Hydrophilic
         <u>Peptide</u> hormone
         Neurotransmitter

   (b) (i) Four ring structure (with variable groups attached)
       (ii) (Affects) fluidity/permeability (of
         membrane/phospholipid bilayer)
       (iii) 6:1:2

6. (a) (Repressor binding to the operator switches) structural
       gene off/not transcribed/not expressed
       or
       no beta galactosidase/lactase/lactose-digesting enzyme made

   (b) (i) Secondary
       (ii) Basic
       (iii) Positive charges (of lysine) interact/bond
         with/attracted to the negative charges on DNA

7. (a) $50\mu Ml^{-1}$ (micromoles per litre); 51 OK

   (b) (Competitive) inhibitor occupies active site **and** a higher
       concentration of/more substrate is needed to reach (half)
       V_{max}/same rate of reaction

   (c) 5×10^{-4}s or other formats

8. **A** 1. glucose exists in linear and ring forms
2. more of the ring form
3. description of C1 OH group in α position below plane/ring and β above plane/ring
4. glucose monomers join by condensation reaction/glycosidic bonds
5. diagram of alpha (1,4) bond **or** beta (1,4) bond **or** alpha (1,6) bond
6. (side) branches are between C1 and C6
7. starch is found in plants **and** function is **energy** storage
8. two forms of starch - amylose and amylopectin
9. starch has glucose monomers joined by α (1,4) bonds
10. amylose linear/has no (side) branches
11. amylose forms a helix
12. glycogen is found in animals **and** function is **energy** storage
13. glycogen has glucose monomers joined by α (1,4) bonds
14. amylopectin has less branching than glycogen
15. cellulose is found in plants **and** provides support/is structural
16. cellulose has glucose monomers joined by β (1,4) bonds
17. cellulose has glucose molecules inverted every second molecule
18. gives a rigid/straight chain
19. chains form fibrils
20. starch and glycogen are insoluble **and** do not affect osmosis

B 1. PCR is used to **amplify** DNA
2. DNA denatured/melted by heating to about 90°C
3. cooled to let primers anneal/attach
4. primers are short base sequence complementary to sample
5. primers delimit the copy region
6. Taq polymerase optimum temperature about 70°C
or
Taq polymerase stable at high temperature
7. polymerase synthesises complementary strand
8. cycle repeated 30 or so/many times
9. used for forensics/paternity disputes/testing pedigrees
10. DNA samples are isolated/purified
11. samples digested with restriction enzyme/endonuclease
12. differences between individuals' genomes
or
reference to hypervariable repeat sequences/VNTR
13. fragment lengths of digested DNA vary between individuals
14. fragments separated by electrophoresis
15. electric current/electricity passed through gel
16. DNA has a negative charge/phosphates charged
17. shorter fragments travel further/faster
18. fragments blotted onto a filter/membrane
19. probe is a short chain of single stranded DNA
20. probe binds to complementary bases on fragments/hybridisation occurs
21. fragments with probe attached show up
or probe is labelled
22. Correct description of outcome

SECTION C: BIOTECHNOLOGY

1. (a) (i) ELISA/immunoassay
(ii) Antibodies are specific to antigens
or
HSV antigens/epitopes not the same as those for chickenpox
(iii) Some antibody R-enzyme would be present and react with substrate
even though it is not bound to Q

or even though no Q is present
or even if some Q is present but unattached (to HSV antigen)

(b) Myeloma and (B) lymphocyte/B-cells

2. (a) (i) Lactose to lactic acid
(ii) Pasteurisation
or
to kill pathogens/naturally occurring bacteria/spoilage bacteria

(b) (i) Two species present/two different types of cell present
(ii) *Any one from*:
• culture has become contaminated
• pasteurisation process not successful
• two bacterial species in added culture

3. *Any 5 points from:*
1. single colony from an agar plate
2. transfer to (nutrient) broth /liquid culture
3. purity/(stage of) growth checked by plating on agar/by microscopy
4. succession of transfers to increasing volumes
5. fermenters for larger volumes
6. and 7. Any two factors for consideration in scaling up
• cost
• containment of micro-organisms
• aseptic transfers
• sterilisation of fermenters
• abiotic – pH/temperature/oxygen/stirring

4. (a) Silo (tower)/bunker/pit

(b) Creates anaerobic conditions (which leads to) **and** lowered pH/acidic conditions/lactic acid
Spoilage organisms do not grow/are killed

(c) *Enterococcus (faecalis)/Lactobacillus (plantarum)*

(d) (i) Lower free amino acids indicates less protein breakdown
(ii) (Added bacteria have) no **significant** effect on lactic acid content
or
If using lactic acid conclusion in d(i) then accept
Significantly lower free amino acids indicates less protein protein breakdown

SECTION C: ANIMAL BEHAVIOUR

1. (a) (i) (Although) Stages 2 and 5 (approx) 25% abundance
or proportion similar
(Birds made) 75% (approx) of foraging choices for stage 2 flowers (0% for stage 5)
(ii) maximising energy gain **and** minimising expenditure
or maximum **net** gain in energy
(iii) Stage 2 flowers have greatest availability of nectar **and** these are visited most frequently
or
As nectar production increases foraging increases

(b) (i) (Energy) cost of gamete production is greater in females
(ii) Genes for nest building have effects outside the body of the bird

(c) Similar/the same

2. (a) (i) Most rapid response is (to silk from) unmated females
(ii) Latency
(iii) Frequency/duration/intensity/strength
(iv) Behaviour may differ from that shown in the wild
Other factors/stimuli from natural habitat not present
Investigator influences behaviour

(b) Prevent/reduce chances of being eaten by females
    or
    Prevent waste of time/energy in courting unreceptive female/increased chance of mating

3. (a) *Any one from*:
   - larger proportion/more likely to be on outside of groups **and** more vulnerable/easier to catch
   - more spaced out/further away from neighbours **and** more vulnerable/easier to catch
   - spend less time scanning **and** less likely to detect cheetah

   (b) 18

   (c) Vigilance

4. 1. agonistic behaviour - all aspects of competitive/fighting behaviour
   2. threat/attack to establish dominance/status (in a social group)
   3. appeasement/submissive behaviour terminates attack/inhibits aggression
   4. example of appeasement/submissive posture eg sexual presentation in baboons
   5. ritualised display allows assessment of opponent/chances of success
   6. (ritualised display) reduces energy expenditure/reduces fighting or risk of injury
   7. example of ritualised display eg roaring/parallel walk in red deer
   8. reduced risk of predation in social group (benefit all group membership)
   9. increased chance of all individuals/subordinates obtaining food

SECTION C: PHYSIOLOGY, HEALTH AND EXERCISE

1. (a) (i) Weight and height
      (ii) 25 – 34
      (iii) Obesity increases with age
      (iv) Hypertension, (C)HD, diabetes (type 2/NIDDM)

   (b) (i) Fat-free mass/lean tissue/muscle conducts electricity
          or
          Fat offers resistance/impedes current
          The higher the impedance/the lower the current **and** the higher the % fat
       (ii) Depends on hydration/kidney function
            Overestimates fat in lean people or underestimates fat in obese people

2. 1. cardiac output (CO) increases
   2. to keep up with demand for oxygen
   3. heart rate and stroke volume increase
      or
      equation $CO = HR \times SV$
   4. CO/SV increase from greater distension of ventricle
      or CO/HR/SV increase from adrenalin
      or CO/HR increase by sympathetic nerves/stimulation
   5. force of contraction/systolic BP/BP increases
   6. general point about redistribution of blood to muscles
      or example of redistribution **from choices below**
      - increased blood flow/vasodilation in skeletal muscle
      - increased blood flow to heart muscle/through coronary arteries
      - increased blood flow/vasodilation in skin
      - reduced blood flow/vasoconstriction in viscera

3. (a) *Any 2 from*:
      Exercising muscles need oxygen for energy release/ATP production/aerobic respiration
      The rate that muscles work at depends on the ability to deliver oxygen to them
      Higher VO_{2max} indicates higher oxygen consumption/higher fitness

(b) (The fitness value is) derived by dividing O_2 uptake by body mass
    or it is **per kg** tissue
    or quantify using eg dividing max O_2 by body mass

(c) **2.896** or **2.9** $l\ min^{-1}$

(d) Patient recovering from MI or heart condition
    Elderly

4. (a) Thermic effect of food/dietary (induced) thermogenesis

   (b) *Any one from*:
       Either sugar increases energy expenditure
       (Energy expenditure) for sucrose more than glucose
       Different sugars give different changes in EE

   (c) *Any one from*:
       - difference between sugars is not due to chance/biological variation
       - significant difference between glucose and sucrose results
       - obtained by testing a number of individuals
       - glucose results more reliable
       - results are mean values
       - there is variation between individuals

   (d) Direct – measures heat output **and** indirect – measures oxygen uptake etc

BIOLOGY ADVANCED HIGHER
2010

SECTION A

1. C	14. C
2. D	15. B
3. C	16. D
4. A	17. B
5. A	18. B
6. C	19. D
7. B	20. C
8. C	21. D
9. A	22. D
10. B	23. A
11. D	24. B
12. D	25. B
13. A	

SECTION B

1. (a) it/oncogene is **dominant**
   the mutation/mutated allele is dominant
   mutated proto-oncogene/mutated proliferation gene are dominant

   (b) (i) monolayer/one layer/single layer
   confluent/**complete** coverage
   (ii) stimulates cell proliferation/division
   or
   provides/is a source of growth factors/MPF

   (c) (i) the two genes/*EML4* and *ALK* alone/not fused/do not transform cells/produce foci
   (ii) modification of fusion (gene) stops foci formation/transformation
   modification removes kinase activity

   (d) (At zero inhibitor conc. in Fig 2 and Fig 3)
   both normal and transformed cells start at 10^6 cells and both reach 10^{10} cells by day 7

   (e) (i) (at day 7, cell number) in absence of inhibitor is 10^{10} and 10^8 in presence of inhibitor **and** difference of $\times 10^2$
   (ii) *Any one from:*
   • as concentration of inhibitor increases, (rate of) growth/proliferation decreases
   • inhibitor reduces growth more in transformed cells than normal cells
   • both 5 µmol l^{-1} and 10 µmol l^{-1} inhibitor reduce cell **number** relative to 0 µmol l^{-1}
   (iii) comparison of final cell numbers **or** of numbers at a specified period/time from both graphs

   (f) *Any one from:*
   This is the concentration that:
   • made transformed cells decrease
   • was effective at killing transformed cells
   • reduced transformed cell number to 10^2
   • would reduce tumour size
   very little effect on normal cells

2. (a) peptidoglycan
   (b) (i) C1 (of NAM) bonded to C4 (of NAG)
   OH of C1 is above plane/ring/carbon
   (ii) hydrolysis

3. *Any five from:*
   1. protein that spans the membrane
   2. works against concentration gradient/by active transport
   3. ATP provides phosphate
   4. phosphate attaches to pump/protein/protein phosphorylated
   5. phosphorylation/dephosphorylation alters conformation/shape of protein **or** description
   6. different conformations have different affinity for sodium/potassium
   7. (3) sodium ions (pumped) out of cell and (2) potassium in

4. (a) contains foreign DNA **or** contains DNA from another **species**
   (b) *Agrobacterium (tumefaciens)*/*A. tumefaciens*
   (c) selects for plants/cells with the plasmid/resistance gene **or** is toxic to (plant) cells with no plasmid/resistance gene surviving (plant) cells now have the desirable/transgene sequence
   (d) bacterial toxin/Bt toxin/insecticide in tomato/other plants
   herbicide resistance added to in corn/maize/cotton/other plants
   beta carotene added to/in rice
   increased shelf life (flavr savr) tomato

5. (a) (In intraspecific, individuals are from the same species so) individuals have the same resource needs/niche
   or
   converse for interspecific
   (b) (i) B
   (ii) any competition is negative
   or
   presence of any competitor reduces time spent feeding/at flower
   or
   interspecific more intense than **intra**specific
   (c) butterfly has other food/nectar sources
   or
   brambles pollinated by other insects /nectar eaten by other insects
   or
   idea that relationship is not intimate/fixed/symbiotic

6. (a) herbivores eat plants **and** detritivores eat dead/waste material
   (b) (i) comparison of sample 1 and sample 4
   at week 6:
   Sample 1: 11-12 % loss (88% remaining)
   Sample 4: 90% loss (10% remaining)
   or
   compare data from sample 1 and sample 4
   sample 1 has 85% left at 12 weeks and sample 4 reached 85% remaining by week 1 (roughly). So rate of decomposition is 12 times faster = > 7
   Single mark options
   If not comparing Samples 1 and 4 but data show >7
   or
   losses calculated for Samples other than 1 and 4
   (ii) error bars/results overlap **and** no difference in results for types of gut/amount of decomposition
   or
   no **significant** difference
   or
   wide error bars so results very variable/are less reliable
   (iii) material has been through two animal guts
   and
   idea that fragmentation/surface area increasing (so this sample has the highest rate of decomposition)

7. (a) with pollution there are favoured and susceptible species

indicator species defined as those that experience consequences/are sensitive to a pollutant

or

example showing **how** a favoured/susceptible species indicates pollution

(b) (i) 11 000 000

(ii) not enough time for diclofenac to reach safe levels or equivalent

8. **A** (i) *Any seven from:*

1. yield reduced by competition/disease/damage/herbivores
2. (any of above) controlled by reducing populations (to maximise food production)
3. weeds are competitors for the crop's resources
4. herbicides used to kill other plants/weeds
5. alternative methods of reducing weeds, e.g. hoeing, interplanting, (herbicide-tolerant) transgenics
6. more resources for the crop (increases yield)
   or e.g. of resource – nitrate, light, space
7. insects (damage crops by) eating plant parts
8. insecticides reduce insect populations/kill insects
   or (Bt) transgenic **or** biological control **or** barrier
9. herbivores removed or kept out (e.g. scarers, netting, fencing)
10. parasites/fungi/viruses reduce yield by causing disease
11. fungicides prevent/kill fungi
12. example of ecological impact of these activities

(ii) *Any eight from:*

13. cultivation of one species/crop (to exclusion of others)
14. to meet the demand/needs of increasing population
15. removal of hedgerows allows increased field size
16. reduces species diversity (e.g. predators)
17. reduces stability of ecosystem
    or increases susceptibility to pathogens
18. mechanisation increases efficiency of cropping/planting
19. reduce food costs/increases profit
20. may damage soil structure/increase compaction
    or may have negative impact on soil condition
21. cultivation of single species depletes particular nutrients
22. need for fertiliser/use of inorganic fertiliser (improves yield)
23. but can cause knock on pollution problems/eg of effect

8. **B** (i) *Any five from:*

1. energy fixed/light to chemical in photosynthesis
2. (photosynthesis carried out) by autotrophs/producers
3. productivity is **rate** of accumulation of biomass
   or productivity is mass units per area per time
4. primary productivity supports higher trophic levels
5. GPP is total yield of organic matter / total energy fixed
6. NPP is biomass remaining after producer respiration
   or equation
7. appropriate management/choice of biomass crop
8. explanation of role of biomass in the energy debate
   or relevant comment about 'carbon footprint'/neutrality

(ii) *Any ten from:*

9. fossil fuels are finite/non-renewable (energy) resources
10. need to be conserved
11. biofuels are alternative sources of energy
12. fossil fuels are **burned**
13. gases released SO_2/ NOx/ CO_2
14. these cause acid rain
15. effect of acid rain in ecosystems (e.g. forest damage, pH in lakes, etc)
16. CO_2 (from combustion) **enhances** greenhouse effect
17. cause global warming/climate change
18. changes abundance or distribution of species (general point)
    or example **other than** zooxanthellae
19. zooxanthellae and polyps/coral are symbiotic/mutualistic
20. and coral bleaching link to temperature rise
21. methane and CFCs other important greenhouse gases not from fossil fuels

Section C: Biotechnology

1. (a) (i) nif/nitrogen-fixing genes
   (ii) nitrogen converted to ammonia/ammonium

   (b) *Any two from:*
   nitrogenase activity declines with length of treatment

   no significant difference between 10 and 20 days

   there is a (significant) decrease in activity from 20 to 30 days

2. *Response to antigens*
   1. antigens are recognised by B-lymphocytes
      or antigens activate B-lymphocytes
   2. multiplication (of B-lymphocytes) in spleen/immune system
   3. antibodies are produced
   4. antibody binds/forms complex with antigen
      or antibodies are specific

   Application
   5. (highly purified) antigen injected into animal
   6. bleeding of animal
      or red blood cells removed from blood sample
   7. reinjection (with antigen) increases response
   8. polyclonal serum produced **or** antibodies prepared from serum

3. (a) *Any one from:*
   Uniformity in terms of:
   • clones/genetic
   • growth rate
   • ripening
   • harvesting time
   • yields
   • taste
   high yielding disease free/virus free/economical

   (b) light intensity/temperature/humidity/nutrients/pH/water content

   (c) (i) treatments are an improvement on the control **or** example
       or
       as concentration of regulators increase the explants increase
       (ii) (At 0·5 IAA + 5·0 kinetin)
       $58 \times 28 = 1624$ shoots = highest yield

4. (a) competition with gut pathogens **or** control pathogens **or** reduce diarrhoea
   anti-cancer activity
   reduction of blood cholesterol
   reduced lactose intolerance

(b) (i) direct count would include living and dead
   or dilution plating gives viable count
   or alternative comparisons, eg **live** bacteria needed so need **viable** count
   (ii) prevent contamination/maintain aseptic technique
   (iii) only count between 30 and 300
      or
      enough to be accurate and not too many to count
   (iv) correct value 6×10^{10}

      plate to bottle 4 = 6×10^3

      dilution factor from bottle 1 to bottle 4 = **10^4**
      ($10^2 \times 10 \times 10$)

      scaling 0·1 cm³ sample from 100 cm³ in carton = 10^3

Plate ——————————> bottles ——————————> carton
60 in 0.1 × 100 = 6000   ×10 ×10 ×100   ×1000

Section C: Animal Behaviour

1. (a) background/height/tree species/type of woodland etc

   (b) B or E

   (c) (B was expected to give best survival but)
      Little difference between B and C
      or C is effective yet has only one spot/has 'no eyes'
      (E with two 'eyes' was expected to give high survival but)
      Increase in number of spots increases survival
      F–E–D (24/48 hrs)
      (A, the control has no contrast and)
      In all treatments survival is higher than control
      B–C–D have highest contrast and highest survival
      Appropriate quantification

   (d) mimicry/batesian mimicry/mullerian mimicry
      crypsis/camouflage
      masquerade
      disruptive coloration
      aposematic/warning coloration

2. *Any four from:*
   1. hierarchy is system of (social) ranking
   2. established by fighting/maintained by threat or ritualised display
   3. reduced aggression/fighting (once established)
   4. increased protection/better chance of survival
   5. co-operative hunting/all group members get food
   6. division of labour/opportunities for learning

3. (a) ethogram

   (b) (i) sexual dimorphism
      (ii) to attract females/to permit female choice/to elicit response from females
   (c) (i) brother-sister matings
         or
         mating (only) between closely related fish
         or Prevent dispersal/keep all family individuals in the same area
      (ii) reduces breeding success **or** less fertilisation and less hatching
         Inbreeding causes:
         increased expression of disadvantageous/lethal/ recessive genes
         increased homozygosity/inbreeding depression

4. (a) (i) high-learning flies die younger/shorter life span
      (ii) (on average/in either group) females live longer than males
         or
         greater reduction in life span/longevity for females (in high-learning group)
      (iii) much longer life span in primates

      or
      learning has bigger role in primate behaviour
      or
      learning benefits outweigh the costs in primates

   (b) *Any two from:*
      irreversible/difficult to reverse
      environmental component/object in environment
      critical time period (after hatching/birth)
      object (of attachment) followed to exclusion of others

Section C: Physiology, Health and Exercise

1. (a) *Any two from:*
      3+ hours/endurance level of exercise results in lower (resting) heart/pulse rate than **the others**

      3+ hours/endurance level of exercise results in greater left ventricle mass than **the others**

      athletic heart defined as increased LV thickness **and** increased SV/lower (resting) pulse

   (b) stroke volume

   (c) increased LV mass

   (d) less plaque/atheroma build up/atherosclerosis
      better lipid profile/higher HDL:LDL ratio/higher HDL/lower LDL
      better myocardial circulation
      lower (resting) BP
      lowers risk of MI/stroke

2. *Any five from:*
   1. pancreas senses glucose (in blood)
   2. insulin secretion/production (increases) when glucose is high
   3. insulin increases glucose uptake by liver/muscle/cells
   4. insulin **promotes** glycogen synthesis/conversion of glucose to glycogen
   5. insulin increases the number of glucose transporters (in cell membranes)
   6. obesity leads to/is a risk factor for NIDDM (Type 2 diabetes)
   7. obesity/NIDDM leads to insulin resistance/loss of sensitivity to insulin/fewer **active** receptors
   8. reduced entry of glucose into cells
   9. (in NIDDM) insulin concentration initially increases

3. (a) bone mass/density is increasing in this period
      or
      to increase bone density (before age-related loss)
      or
      weight bearing activity is likely to be high

   (b) reference to menopause around this age/oestrogen starts to decline/women start to lose bone mass rapidly/when osteoporosis (most likely) starts

      may counteract/slow down (age-related/oestrogen-related) losses
      or
      maintain bone density/mass
      or
      delay onset of osteoporosis

4. (a) (i) measures heat output/loss

         in **insulated** chamber/environment
         or measure energy required to keep chamber temperature constant
      (ii) volume of air breathed
         % oxygen breathed in and % of oxygen in air out
         or difference in percentages

   (b) (i) 838.3 kJ
      (ii) select foods with biggest deviation from 20·20 kJ (starch or protein)

correct calculation of either extreme (as fraction of single food value)

starch (low) = 4·62%   protein (high) = 4·87%

or
5% limits from 20.20 are 19·19 kJ to 21·21 kJ
and
conclude all four are in range

or
4 deviations calculated correctly
and
conclude all four are in range

BIOLOGY ADVANCED HIGHER 2011

SECTION A

1.	C	14.	A
2.	B	15.	A
3.	B	16.	C
4.	D	17.	C
5.	D	18.	D
6.	B	19.	C
7.	A	20.	D
8.	A	21.	C
9.	B	22.	B
10.	D	23.	D
11.	D	24.	B
12.	A	25.	B
13.	C		

SECTION B

1. (a) (i) *Any two from:*
        Type of symbiosis **or** idea of close/intimate association between two **species**
        <u>Host</u> harmed and <u>parasite</u> benefits
        Benefit (to parasite) in terms of nutrition/energy/resources
    (ii) Obligate

    (b) (i) Idea of checking human faeces (for parasite eggs)
    (ii) *Any two from:*
        Health education/ideas on how to reduce infection
        Sanitation/prevent faeces reaching lake
        Drug treatment (for superspreaders)

    (c) (i) Intervention village(s) reach target/1% and control(s) don't.
        Use data to illustrate trends
    (ii) Commit to position about the results (reliable or unreliable) **and** justify appropriately, eg 'Reliable because two villages used for treatment and control', or 'not reliable because (*only*) *two* villages …' Reliability in relation to **error bars** – when error bars are small the reliability is better

    (d) (i) Difficult to kill them all
        or
        Survivors reproduce rapidly
    (ii) *Any one from:*
        (Parasites still exist as)
        Adults long lived/still inside host
        Eggs still being produced
        Free-living parasite stages not affected

    (e) *S.japonicum* can infect other **mammals**/can have a range of primary hosts
        (so) cattle have to be kept away from lake
        (and) mice can be used for test purposes

2. (a) photosynthesis **and** respiration

    (b) 12.5%

    (c) (i) added to by human activity
    (ii) methane/CFC/nitrous oxide/ozone

3. (a) (i) earlier species **change conditions/environment** to better suit later species
    (ii) allogenic

    (b) (i) bioaccumulation/bioconcentration
    (ii) high toxicity so herbivores/primary consumers die

or

biomagnification makes herbivores too toxic (for carnivores)/biomagnification results in toxicity in higher trophic levels

or

low productivity/too little energy to support higher levels

(iii) survives high conc. of nickel while **other species** susceptible

or

the **relative abundance** depends on Ni levels of soil

4. *Any four from:*
   1. Fundamental niche is the resources a species is capable of using/could use in the absence of competition
   2. Realised niche is the resources a species actually uses or available in presence of competitors
   3. (Competition arises) when resources limited
   4. Competitive exclusion arises from interspecific competition/when two **species** competing
   5. Two **species** with the same/similar niche cannot coexist (in same location)
   6. (One species will survive and) one species will die out/local extinction

5. (a) (i) (For the increase in O_2 pressure 0–30 units)
   Myoglobin increases to 0.975
   Haemoglobin increases to 0.50
   (ii) curves differ/binding differs
   tertiary (structure) the same/similar
   only Hb has quaternary

   (b) 1 (less)

   (c) prosthetic groups

6. (a) (i) hydrophilic/not lipid soluble
   (ii) (signal) transduction

   (b) (i) at most/all GABA concentrations more chloride movement (with drug present)
   (ii) change in conformation/shape (of the GABA receptor)

7. (a) single-stranded DNA
   (bases) complementary **or** strand anneals (to template)

   (b) *Any one from:*
   (gene) probes/probing
   (gel) electrophoresis
   blotting
   sequencing
   restriction digest

   (c) (Test is negative for ΔF508 so counselling needs to warn of) other possible mutations (30%) causing CF
   or low chance of having/carrying CF

8. **A** (i) *Any five from:*
   Prokaryotic DNA
   1. within cytoplasm/not contained in a nuclear membrane
   2. exists as a circular DNA molecule/nucleoid
   3. plasmids are **additional** circles/rings of DNA
   Eukaryotic
   4. contained within a nuclear membrane
   5. DNA is associated with histone/proteins
   6. organised as nucleosomes/chromatin
   7. (nucleosomes) coiled/condensed to form chromosomes
   8. chromosomes are linear

   (ii) *Any ten from:*
   Prokaryotic
   9. ribosomes (only organelle)
   10. cell wall made of peptidoglycan
   11. capsule/layer of mucus (lipopolysaccharide) is protective/is adhesive
   12. pili for cell attachment/exchanging plasmids
   13. flagella for movement
   Eukaryotic
   14. name and function of one organelle from list below
   15. name and function of another organelle from list below
   16. cytoskeleton is a system of protein fibres that provide support **or** movement **or** movement/organisation of organelles
   17. animal cells (may) have microvilli to increase surface area/absorption
   18. plant cell walls made of cellulose
   19. middle lamella is where plant cell walls contact (rich in pectin)
   20. plasmodesmata connect cytoplasms/adjacent plant cells
   21. plant cells (may) also contain – chloroplasts for photosynthesis **or** vacuoles for cell sap

8. **B** **Eukaryotic**
   Endoplasmic reticulum – transport of proteins/synthesis of lipids
   Golgi apparatus – processing/modification/secretion of proteins
   Mitochondrion – (aerobic) respiration/ATP production
   Lysosomes – enzymatic digestion
   Microbodies/peroxisomes – oxidation reactions
   Ribosomes – protein synthesis
   Nucleolus – ribosome formation

   (i) *Any five from:*
   1. interphase is G_1, S, G_2
   or interphase is the period between cell divisions
   2. G_1 and G_2 are growth periods **or** organelles/ proteins made
   3. DNA replication occurs during S phase
   4. G_1 checkpoint assesses cell size/mass
   5. G_1 checkpoint ensures there is sufficient (mass) to make two daughter cells/to enter S phase
   6. G_2 checkpoint assesses DNA replication
   7. G_2 checkpoint controls entry into mitosis
   8. ensuring each daughter cell receives a complete genome/'set' of DNA

   (ii) *Any five from:*
   9. spindle fibres are microtubules
   10. correct description of one phase of mitosis – as in notes
   11. as above
   12. M/metaphase checkpoint controls entry to **anaphase**
   13. ensures chromosomes are aligned correctly (on the equator)
   or ensures each daughter cell receives correct number of chromosomes/chromatids
   14. mitosis promoting factor (MPF) needed for entry to mitosis
   or MPF is a protein
   15. cytokinesis is the division of the cytoplasm/separation into two cells

    (iii) *Any five from:*
16. proto-oncogenes/proliferation genes stimulate cell division
17. proto-oncogenes mutate to oncogenes
18. oncogenes stimulate excessive/abnormal cell division/tumour formation
19. tumour suppressor genes/anti-proliferation genes inhibit cell division
20. tumour suppressors act at checkpoints
21. (tumour suppressor) mutation results in loss of inhibition/loss of control of division
22. oncogenes are dominant and in tumour suppressor genes, mutations are recessive

    or

    only single oncogene mutation required whereas two tumour suppressor mutations required

Section C: Biotechnology

1. (*a*) Antibiotic **and** type of organism
   e.g. penicillin and *Penicillium*/fungus
   or
   Correct antibiotic and *Streptomyces*/bacterium

   (*b*) (i) So that only a single species/strain is used to prepare the inoculums
       or idea of pure culture
       or uncontaminated culture
   (ii) Area X

   (*c*) dissolve oxygen/aerate
   or to achieve distribution of nutrients/fungal cells/heat energy/efficient mixing

   (*d*) *Any one from:*
   filtration/ultrafiltration
   addition of salt to a penicillin rich solvent
   precipitation from solvent/flocculation
   centrifugation
   crystallisation

   (*e*) production starts as glucose is (nearly) exhausted
   or lag period 0-1.5 days before production begins

   production begins towards end of active growth/ exponential phase
   or production begins as stationary phase is entered/growth plateaus

2. (*a*) *Any two from:*
   Mouse injected with antigens
   Production of B-cells triggered/activated
   (B-cells) isolated from spleen

   (*b*) Polyethylene glycol/PEG

   (*c*) *Any one from:*
   • mAbs bind to cancer cell-specific antigens
   • immune response against target cancer cell triggered
       or body destroys its own cancer cells
   • delivery of radiation directly to tumours (radioactive molecule can be attached to mAb)
   • delivery of attached toxin to destroy cancer cell
   • treatment of breast cancer using herceptin
   • mAb can prevent growth of cancer cells (by blocking growth receptors)

3. 1. cell walls reduce yield
   2. composition/component: pectin, cellulose, araban
   3. cellulose tough/causes difficulty with breaking open cells
       or cellulose makes mechanical extraction difficult
   4. pectin increases viscosity/causes difficulty with filtration
   5. pectin/araban cause haze/cloudiness
   6. first example of enzyme used to break down the wall materials (see list below)
   7. second example (see list below)
   8. low solubility issues of araban and pectin

   Examples
   Cellulose breaks down cellulose/increases yield
   Pectinase breaks down pectin/decreases viscosity/decreases haze
   Arabanase breaks down araban/decreases haze

4. (*a*) Living cells only

   (*b*) (i) 1.25% and 2.5%
   (ii) same initial concentration of colony forming units/same viable count added to each dilution
   (iii) 9 million cells

Section C: Animal Behaviour

1. (*a*) (i) 15
   (ii) *Any one from:*
   they break more easily/less (total) height needed to break them
   it takes fewer drops to break them
   shorter handling time
   less time/energy to break shells

   (*b*) optimal foraging maximises net energy gain

   it gives the lowest **total** height needed to break a whelk
   the least energy expenditure in flight at this height

   or

   (*c*) Encounter rate (of prey by predator)/search time

2. 1. nature = behaviour that is innate/instinct/genetically determined
       or
   nature allows stereotyped response to stimuli
   2. nurture defined as behavioural modification/learning
   3. nature eg: any example of instinctive behaviour
   4. nurture eg: imprinting/habituation/cultural transmission
       or description of species and behaviour
   5. (adult) invertebrates generally have a shorter **lifespan** than primates *
   6. long lifespan gives time for learning
   7. short lifespan entails reliance on innate behaviour
       or
   invertebrates rely on innate behaviour
   8. invertebrate **parental care** is rare *
   9. primates rely more on **nurture than do invertebrates**
   * converse applies

3. (*a*) (i) *Any two from:*
   Healthy females produce many eggs
   Brood pouch filled faster/reduced mating time
   Reduced predation risks
   Increase in number of eggs fertilised
   (ii) *Any one from:*
   Genes allow more copies to pass into next generation
   Genes more likely to be passed on to next generation
   Genes are self-preserving
   Genes assist survival of the male fish

(b) *Any one from:*
Nutrition of young in brood pouch/carrying young
Providing parental care

(c) Males do not avoid other males with black spots

(d) Fish with solvent only

(e) Males are not influenced by displays/stimuli that females might show if they saw the males

4. (a) (On average) they share half of their genes/genetic material
or The chances of sharing a gene are 0.5/50%

(b) (i) Genes for altruism will spread when rB–C>0
or
helping relatives is beneficial when rB–C>0

(ii) (Three) groups of most related have highest cannibalism
or
no correlation

Section C: Physiology, Health and Exercise

1. 1. deposition of fatty materials/plaque forms/atheroma forms
   2. (atheroma) **under** lining layer/endothelium/intima **of artery**
   3. platelets attach to rough surface/platelets release clotting factors
      or thrombus/clot forms at site of plaque
   4. clot/embolus/atheroma can block/narrow vessel
   5. **blockage** of coronary artery
   6. heart muscle cells **die** beyond blockage
      or heart muscle cells **die** from lack of oxygen

2. (a) **increase** exercise
      reduce intake of fatty foods

   (b) (i) 5.9 (mmol/l)
      (ii) *Any two from:*
         LDL has been reduced (to 2.9)/LDL now within normal range
         Total cholesterol is reduced/now about normal
         Total cholesterol/HDL ratio reduced/now about normal
      (iii) 32.7% **or** (about) 33%

3. (a) pancreas/islets/Beta cells **detect** glucose **and** insulin secretion (increases)

      (Glucose level is reduced when)
      cells in liver/muscle/adipose tissue (increase) uptake glucose
      or
      glucose is converted to/stored as glycogen

   (b) (i) fewer receptors active/functioning/responding to insulin
         or
         receptors do not recruit glucose transporters to the membrane
      (ii) obesity is cause (of insulin resistance/Type 2 diabetes)

         high W:H/this ratio is an indicator for obesity
         (so worth reducing it)
         or
         (reducing ratio) will reduce obesity/BMI

4. (a) (Sporting activities) increase bone mass/bone density/bone mineral density (BMD)
      or
      osteoporosis takes longer to develop because BMD is higher

      greatest bone mass achieved when young/by age of 30/in adolescence
      or
      gives higher BMD before age-related loss

(b) (i) These are most common fracture sites in elderly/those with osteoporosis
   (ii) not a weight bearing exercise and allows comparison with the others
      or
      to demonstrate that **only** weight bearing exercise is effective
   (iii) (Sample data eg)
      size of sample, replication
      variation in BMD/age between subjects
      idea of measuring error, eg error bars

BIOLOGY ADVANCED HIGHER 2012

SECTION A

1.	B	14.	A
2.	D	15.	B
3.	D	16.	A
4.	B	17.	B
5.	D	18.	C
6.	C	19.	D
7.	A	20.	C
8.	C	21.	B
9.	A	22.	D
10.	B	23.	A
11.	C	24.	C
12.	D	25.	A
13.	C		

SECTION B

1. (a) (i) phosphodiester (bonds)/phosphoester
      (ii) (complementary) base pairing **and**
           stops the shape unravelling/creates the shape/holds shape

   (b) *Any two from:*
       Drosha not working
       miRNA/precursor not processed/cut
       no (micro)RNA strand for RISC
       or RISC can't bind (m)RNA
       (RNA) interference reduced/translation is left on

   (c) (i) 62.5
       (ii) more KO cells in G1 **and** fewer in S (and G2+M)
            differences are significant (only) in G1 and S/error bars don't overlap in G1 and S
            or if comparing only G1 bars or only S bars, then must point out significant difference (for 1 mark)

   (d) (i) *Any two from:*
           • in KO cells it is (generally) lower than normal cells
           • it increases in normal cells (over time)
           • in KO cells **and** one from below
               no trend
               decreases from day 8
               increases (to day 8) then decreases
       (ii) in normal cells, as differentiation increases self-renewal decreases
            or converse
            Any one from:
            in KO/abnormal cells, **both** processes decrease after day 8
            or
            in KO/abnormal cells, **both** processes increase to day 8
            or
            in KO abnormal cells, self-renewal remains higher and differentiation remains lower **than normal**

   (e) lactose absent
       repressor binds to operator

2. (a) (i) binding is extracellular/to cell surface/to membrane
            change in cAMP is inside the cell / intracellular
            or
            change in intracellular signal molecule/second messenger
       (ii) crypsis/camouflage/description of camouflage

   (b) (i) (mechanical) support/strength/shape
           cell movement
           spindle fibres/separation of chromatids/movement of chromosomes

       (ii) tubulin
       (iii) centrosome/centriole/MTOC

3. *Any five from:*
   1. inhibitors reduce enzyme activity
   2. competitive inhibitors resemble substrate
   3. (CI) binds to/occupies active site **and** prevents substrate access
   4. non-competitive inhibitors/negative modulators bind at a second site/allosteric site
   5. non CI/negative modulator alters the shape of the **active site**
   6. substrate binding reduced/prevented
   7. idea of modulation (of rate) by change in **affinity** of active site for substrate
   8. allosteric enzymes involved in regulation of pathways

4. (a) probe

   (b) Yes. fragment from P is shorter because of the deletion
       shorter fragment goes further (in the gel)
       or
       DNA with deletion goes further (in the gel)

   (c) polymerase chain reaction/PCR

5. (a) bioaccumulation

   (b) (i) 30% (or 29.64 or 29.6)
       (ii) run-off from land/leaching/spray **and** reaches **sea**
            through food chain/biomagnifications

6. (a) **energy loss** (from food chain) **and** via:
       high metabolic rate/homeothermy/heat/respiration/movement
       herbivory/food source high in cellulose/uneaten parts/undigested parts

   (b) (i) less predation **or** more feeding
       (ii) (further global warming because)
            shorter winter ice duration reduces krill population density
            so less faeces to trap carbon (dioxide) from the atmosphere
            or
            faeces (now) decompose **or** decomposition begins
            carbon dioxide returned to atmosphere

7. (a) suspended growth/suspended development/suspended life cycle/**reduced** metabolism

   (b) (large fields have) fewer hedges/fewer rose plants **so** fewer insects/disrupts life-cycle
       (increased yield from) less feeding/less disease

   (c) organism/species that transfers **parasite** between **hosts**

8. **A** 1. definition of niche: (multi-dimensional summary of) resources/requirements of species
        2. fundamental niche is the resources a species is **capable** of using/could use
        3. realised niche is the resources a species actually uses **or** are available in presence of competitors
        4. competition defined as two organisms attempting to utilise same resource
        5. (competition arises) when resources limited
        6. competition is negative for both species
           or
           competition is a negative-negative/($-/-$) interaction
        7. example of negative outcome/cost (reduced growth, fecundity, population decrease, increased mortality)
        8. intraspecific same species and interspecific different species
        9. intraspecific competition more intense (than interspecific)
        10. because all require same/similar resources

11. exploitation competition defined as use of resource reducing the supply to others
12. interference competition is when access to resource is prevented
13. example of either of above
14. two **species** with the same/similar niche cannot coexist (in same location)
15. competitive exclusion arises from interspecific competition/when two **species** competing
16. (one species will survive and) one species will die out/local extinction
17. resource partitioning allows exploitation/sharing of different components of a resource
18. resource partitioning reduces competition
19. example of resource partitioning
20. definition of exotic species as introduced /alien
21. invasive in terms of better competitor/ lacking predators/ lacking herbivores/lacking parasites
22. example of ecological damage caused by named invasive species

8. B (i) Decomposition
1. decomposition is break down of organic matter to inorganic
   or decomposition is mineralisation
2. soil organisms particularly important
3. detritivores are invertebrates/**or** example
4. ...that produce humus/that fragment detritus/wastes
5. increases surface area so speeds up decomposition
6. enzymes/digestion internal (in detritivores)
7. decomposers are bacteria and fungi
8. enzymes / digestion external (from decomposers)
9. decomposer respiration is final releaser of CO_2
10. decomposition limited by available nitrogen

(ii) Nutrient cycling
A maximum of ten from:
11. finite supply of nutrients
12. cycles maintain the supply
13. uptake/fixation by plants
14. *any two from:* assimilation, transformation, food chains, and decomposition
15. loss of nutrient from ecosystem eg. via leaching
16. low solubility of P limits phosphorus cycle
17. decomposition of (organic) N compounds produces ammonium
    or decomposition is called ammonification
18. nitrification is ammonium to nitrite to nitrate
19. *Nitrosomonas* converts ammonium to nitrite
20. *Nitrobacter* converts nitrite to nitrate
21. denitrification is nitrate to N gas
22. anaerobic / occurs rapidly in waterlogged soil
23. N fixation: nitrogen gas converted to ammonium
24. Cyanobacteria (N fixers) are free-living/in soil/in water
25. *Rhizobium* in root nodules / legumes
26. nitrogenase (that does the N fixing) is inhibited by oxygen
27. leghaemoglobin in nodule reduces oxygen level

Section C: Biotechnology
1. (*a*) spleen

   (*b*) to make the cell line immortal/so that fused cells can divide indefinitely/because myeloma cells are immortal

   (*c*) (i) the medium kills them
       (ii) (unfused lymphocytes) divide a number of times and die/are not immortal/have a limited lifespan

   (*d*) 1st step selects for desired antibody
       2nd step: selection of the cell that makes the (desired) antibody

2. *Any five from:*
   1. plot growth curve/obtain values of cell numbers over specified time period
   2. to identify exponential phase
   3. use exponential phase to calculate g
   4. g = time for population to double/time for one division
   5. growth rate constant (k) = ln2/g
   6. k is number of doublings/generations per unit time
   7. k is used to optimise growing conditions
   8. to maximise enzyme production

3. (*a*) (i) cellulase/pectinase
       (ii) improves clarity/reduces haze/breaks down hemicellulose/removes araban

   (*b*) (i) (affinity) chromatography
       (ii) the **shape** of the enzyme/active site is specific to (the shape of) the substrate

4. (*a*) kills (the majority of the) naturally occurring **bacteria**/ spoilage **bacteria**/harmful **bacteria**/pathogens

   (*b*) (i) lactose to lactic acid
       (ii) generates flavour/texture

   (*c*) (i) 10×10^7 **or** 10^8
       (ii) mixed culture grew to 9×10^7, pure to 3×10^7
       or other correct quantification

Section C: Animal Behaviour
1. (*a*) head up/scanning/not drinking or eating

   (*b*) (i) randomly selected individuals
       focal sampling (for eg 5 mins)/other sampling strategies
       ethogram/check list
       avoid influence by observer/use camouflage, hide etc.
       video recording/use of camera/ remote telemetry
       (ii) *Any two from:*
       vigilance does decrease as group size increases when lions are present (but)
       when lions are absent the vigilance is not related to group size
       lions increase vigilance in smaller groups sizes but not larger (beyond 11)
       (iii) lions present graph depends on one sample of group size 16
       or
       some group sizes have no values (8,9,10,13,14,15)
       or
       some group sizes have only 1 sample (6,7,16)
       or
       some group sizes have large variation in replicates (3,4)

2. (*a*) as body size increases so does crater diameter/size

   (*b*) (i) individually characteristic/highly repeatable/outside body/innate/gene expression determines outward behaviour
       (ii) any built structure; eg nests/homes/traps etc
       or
       behaviours such as herding/shoaling/bird song

   (*c*) crater building/coloration/display behaviour

3. (*a*) (i) younger subordinates groomed more frequently
       or
       as age of subordinate increases, grooming decreases
       (ii) duration/intensity

   (*b*) *Any two from:*
       reinforce close relationships or alliances/develop bonds/ lower dominance threat/maintain social rank/hygiene/ courtship

(c) "… **not necessarily closely related** to the young animals that they help to rear."

they are not assisting the survival of their own genes

or

altruism generally involves kin selection/close relatives (and this isn't)

4. *Any five from:*

1. sign stimuli/releasers are signals that elicit **specific** response
2. FAP is automatic/stereotyped/species specific response
3. (FAPs are) under genetic control/innate
4. (FAPs are) resistant to change (by learning)
   or once initiated go to completion
5. series of releasers/FAPs can produce complex behaviour
6. (herring) gull adult has red spot on bill (releaser)
7. elicits (FAP of) chick pecking at spot
8. (then releases) parent provides food

Section C: Physiology, Health and Exercise

1. (a) systolic = 120 and diastolic = 70

   or

   systolic/diastolic

   or

   systolic and diastolic pressures in mmHg

   (b) narrowing/obstruction/loss of elasticity increases resistance

   or

   force on artery wall increases pressure

   or

   to keep the same flow (rate) the BP goes up

   (c) (i) 40.7%

   (ii) (with exertion) heart's demand for oxygen increases/ heart rate increases

   but

   diastole shorter

   or time for O_2 delivery shorter

   or lower coronary circulation/oxygen to cardiac muscles

   (d) 5 **or** 10mg **significantly** better than control

2. (a) not obese. BMI = 27 **and** less than 30/obesity cut-off

   (b) (i) mass and volume

   (ii) used in (Siri) equation   **or**   (495/density) − 450

   (c) underestimates fat

3. (a) *Any one from:*

   improve increase HDL:LDL

   improve/increase HDL

   reduces LDL / reduces triglycerides / reduces cholesterol

   (b) *Any two from:*

   increases glucose uptake (in muscle and fat cells)

   increases number of **active** receptors

   increases number of glucose transporters (in target cells)

   increases (enzymes for) glycogen synthesis

   (c) starting/fasting insulin concentration is reduced by exercise **and** correct quantification

   or

   starting/fasting insulin concentration is reduced equally at both exercise levels **and** correct quantification

   or (in response to same food intake) both exercise levels result in lower insulin production than control **and** correct quantification

   or

   insulin increases less after exercise **and** correct quantification

   or

   as the level of exercise increased the insulin response to the meal decreased **and** correct quantification

4. *Any four from:*

   1. individual overweight because energy in > energy out
   2. exercise increases energy out/gives negative energy balance
   3. exercise leads to weight loss/fat loss
   4. exercise (may not alter weight) may increase muscle/alter balance between muscle and fat
   5. BMR increase causes increased energy output
      or
      more lean tissue so higher BMR
   6. the effect of exercise on body mass may decrease as fat decreases
   7. reference to exercise programme: (*any one from*) frequency, duration, intensity and type of exercise

Hey! I've done it

iBrightRED
PUBLISHING

© 2012 SQA/Bright Red Publishing Ltd, All Rights Reserved
Published by Bright Red Publishing Ltd, 6 Stafford Street, Edinburgh, EH3 7AU
Tel: 0131 220 5804, Fax: 0131 220 6710, enquiries: sales@brightredpublishing.co.uk,
www.brightredpublishing.co.uk

Official SQA answers to 978-1-84948-300-1
2008-2012